Microeconomic Impacts of Institutional Change in Vietnam's Northern Uplands

DEVELOPMENT ECONOMICS AND POLICY

Series edited by Franz Heidhues, Joachim von Braun and Manfred Zeller

Vol. 71

PL ACADEMIC RESEARCH

Microeconomic Impacts of Institutional Change in Vietnam's Northern Uplands

Empirical Studies on Social Capital, Land and Credit Institutions

Camille Saint-Macary

PL ACADEMIC RESEARCH

Bibliographic Information published by the Deutsche Nationalbibliothek
The Deutsche Nationalbibliothek lists this publication in the Deutsche Nationalbibliografie; detailed bibliographic data is available in the internet at http://dnb.d-nb.de.

Zugl.: Hohenheim, Univ., Diss., 2012

Library of Congress Cataloging-in-Publication Data
Saint-Macary, Camille, 1982- author.
Microeconomic impacts of institutional change in Vietnam's northern uplands empirical studies on social capital, land and credit institutions / Camille Saint-Macary.
pages cm. — (Development economics and policy, ISSN 0948-1338 ; vol. 71)
Includes bibliographical references.
ISBN 978-3-631-63949-8 — ISBN 978-3-653-04253-5 (e-book) 1. Vietnam, Northern—Economic conditions. 2. Vietnam, Northern—Rural conditions. 3. Uplands—Economic aspects—Vietnam, Northern. 4. Land tenure—Vietnam, Northern. 5. Agricultural credit—Vietnam, Northern. 6. Infrastructure (Economics)—Vietnam, Northern. I. Title.
HC444.Z7S25 2014
330.9597'1—dc23

2014016205

D 100
ISSN 0948-1338
ISBN 978-3-631-63949-8 (Print)
E-ISBN 978-3-653-04253-5 (E-Book)
DOI 10.3726/978-3-653-04253-5

PL Academic Research is an Imprint of Peter Lang GmbH.
Peter Lang – Frankfurt am Main · Bern · Bruxelles · New York · Oxford · Warszawa · Wien

This publication has been peer reviewed.

www.peterlang.com

Editors' Foreword

Vietnam has experienced considerable economic growth and an impressive reduction in poverty over the past two decades. These successes are attributed to the reform program *Doi Moi*, which entailed a gradual transformation of institutions, and a redefinition of the roles of the state, markets and communities in allocating resources. Agricultural markets were gradually liberalized, user rights were transferred to smallholder farmers for most agricultural land, anti-poverty programs were implemented, and further reforms were undertaken to improve rural households' access to financial services. The socialistic state apparatus is still present in rural life via the omnipresence of mass organizations, such as farmers' unions, and women's unions at the commune and village levels. These mass organizations remain controlled by the party and by the state, and are an important element of social capital in Vietnamese communities.

This thesis is concerned with the development gap observed between lowland and upland areas, and the question of why reforms that have been so successful nationally have not proven as effective in mountainous regions. Indeed, mountainous regions and their inhabitants have lagged behind in the process: the poverty incidence in the Northern Uplands in 2008 was two times that of the rest of the country. Furthermore, income inequalities between lowland and upland areas, and between ethnic minorities and the Kinh majority, have risen. The combination of poverty and the degradation of natural resources remains a primary concern in the area, begging the overall policy question of whether institutional reforms in Vietnam can effectively contribute to the reduction of poverty and to more sustainable land use in mountainous regions.

Drawing on a conceptual framework that highlights the determinant role of institutions in the poverty-environment nexus, which is a set of complex linkages between natural resources and livelihoods, this thesis investigates to what extent the current institutional framework addresses objectives of equity, economic growth and environmental sustainability in a mountainous district. It focuses on three critical dimensions: the definition of property rights, the functioning of intertemporal markets, and social capital within communities. Analyses build on a rich and primary quantitative household- and village-level dataset collected in 2007/2008 in Yen Chau, a mountainous district of the Northern Upland region, as well as on qualitative information gathered in the field through focus group discussions and informal discussions with local stakeholders. The author spent about one year in Vietnam conducting the field work.

The clear formulation of research questions and hypotheses, and the high analytical rigor with which they are tested enable the author to make substantial contributions to the literature with this thesis. She identifies sources of success and failure in the current institutional framework to promote sustainable development in Vietnam's mountainous areas and derives policy recommendations there from. Evidence highlights the limitations of the top-down approach that dominates public intervention in mountainous areas, and the omnipresence of the state that impedes the functioning of land and financial markets. Moreover, these interventions are costly and not always successful in enhancing efficiency or fostering environmental sustainability in resource use. This stresses the need for the Vietnamese government to further enhance the functioning of incentive-based mechanisms in the economy as a complement to current policies, and to support the emergence of an independent civil society, enabling ethnic minorities to participate in the public debate.

Acknowledgements

This book is the outcome of my PhD research conducted at the Department of Rural Development Theory and Policy of the Institute of Agricultural Economics and Social Sciences in the Tropics and Subtropics the University of Hohenheim. This research has been realized within the special research program SFB 564, known as the Uplands Program, and financed by the DFG whom is gratefully acknowledged. Working in such a stimulating project under the supervision of Prof. Zeller has been a fantastic experience. Many thanks go to him, and to Prof. Lippert, and Prof. Birner for taking part in my PhD committee and reviewing my thesis. I also thank all the persons in the project and outside who have contributed intellectually and administratively to the achievement of this work, in particular Prof. Heidhues, Alwin Keil, Prof. Stahr, Gerhard Clemens, Holger Frölich, Mrs. Hong and many more. The dataset which constitutes a core part of this thesis has been collected in collaboration with the Department of Economics and Rural Development of the Hanoi University of Agriculture, Prof. Pham Thi My Dung, Le Thi Ai Van and Dinh Thi Tuyet Van. This thesis owes a lot to the enumerators, local officials and respondents in Yen Chau, who have taken part in the survey and provided the main material of this thesis. Many more people, colleagues, friends, and family, have accompanied me through this journey and have contributed in some way to this book, to all of them, thank you.

Executive Summary

Vietnam's economic achievement over the last two decades constitutes one of the most successful development stories of the last century. The *Doi Moi* reform program voted in 1988 gradually led the transition from a centrally-planned to a market-oriented system through the liberalization of markets, the individualization of land access, among other reforms. It has permitted high economic growth rates in all sectors of the economy, and led to massive reduction of poverty in the country. Yet, successes have not taken place evenly through the country. Mountainous regions and ethnic minority groups have lagged behind in this process, and the poverty incidence in the Northern Uplands was still twice as large in 2008 than in the rest of the country (31.6% against 14.5% for the whole country). There in addition, the rapid population growth combined with the intensification of agricultural systems in fragile ecosystems have considerably increased pressure on natural resources, and thereby accentuated the risk for the economy to get trapped in a vicious circle whereby environmental degradation and poverty mutually reinforce each other and durably undermine economic development. The poverty-environment nexus or the set of complex linkages relating poverty and the environment imply that in fragile ecosystems, equity, economic growth and environmental sustainability can no longer be treated as separate objectives but must be jointly addressed in order to enable sustainable development in mountainous areas.

Drawing on a conceptual framework that highlights the role of institutions in the poverty-environment nexus, this thesis examines the sources of success and failure in the current institutional framework with respect to these three objectives. It focuses on three aspects that are considered essential for sustainable agricultural development, namely (*i*) the definition of property rights, (*ii*) the functioning of intertemporal markets such as credit market, and (*iii*) the amount of social capital that communities and households can build on to manage collective resources and overcome market failures. Each of these aspects are investigated empirically in the three articles and complementary chapter composing this thesis. Empirical evidence builds on a rich and primary quantitative household- and village-level dataset collected in 2007/2008 in Yen Chau, a mountainous district of the Northern Upland region, as well as qualitative information gathered in the field through focus group discussions and numerous informal discussions with local stakeholder.

The first article examines the land reform voted in 1993, which after forty years of collectivization was aimed at privatizing and securing land access through the issuance of long term use right certificates. The objective of this large titling program is to boost agricultural production, enhance the conservation of natural resources and reduce poverty. After analyzing the history and content of the land law, the article describes the implementation of the reform in Yen Chau. It then investigates to which extent the reform is influencing the adoption of soil conservation technologies by farmers through descriptive statistics and econometric models of agroforestry adoption that account for selection bias due to incomplete exposure. Results first reveal that the implementation of the reform has been a long and costly process, particularly in mountainous regions where the overlap of the law with customary land institutions has been a source of conflicts. The necessity to enforce the law, but also to keep control over land use, and to ensure the equity in land distribution, led the local government to reallocate lands that had already been distributed to farmers. A question explored in the article is therefore whether these reallocations affect farmers' trust in land institutions, and thereby influence their incentives to adopt long term conservation technologies such as agroforestry. Results show that the possession of a formal land title positively influences adoption, but that the threats of land reallocations in villages discourage conservation practices by creating uncertainty and tenure insecurity. Further analyses reveal that these two effects interact, and thus that the land policy and the way it is implemented affect farmers' conservation practices. One contribution of the article to the literature is therefore to show that the credibility of the state matters and affects farmers' anticipations and response to policies. An other important finding is that beyond knowledge constraint and under the current institutional framework, the conservation technologies currently available in the area remain economically unattractive. Farmers are unlikely to bear the costs for conservation alone. The public good character of soil conservation calls for a greater involvement of research and government agencies to identify land use options that are both economically attractive and environmentally sustainable, and enhance institutional mechanism to foster their adoption (through payments for environmental services for instance).

This article is followed by a complementary chapter that investigates whether the land reform has led to the establishment of a land market in Yen Chau. Well-functioning land markets are expected to generate important equity and efficiency gains in the agricultural sector, by enabling the transfer of land from the less productive to the more productive, and by raising farmers' asset value. Studies showed that this has been the case at the national level, and that the sale and lease transactions taking place in the land market lead to a greater efficiency in land use and a more equitable distribution. Our data shows that the situation in Yen Chau is otherwise. There, land sale transactions remain anecdotal, and the lease market is very thin. Most of the land transactions observed occur through intra-familial free transfers. Results show that these transactions induce a more equal distribution than the initial land allocation. The administrative red tape, the tenure insecurity due to land reallocations and the uncertainty regarding the extension or reallocation of use rights at the end of their term are seen as the main factors hindering the land market. The credit constraint a does not seem to play a major role here.

The second article of this thesis examines the credit market, and investigates the role and impact of state interventions in inducing a more equitable and efficient allocation. The functioning of credit markets plays a crucial role in the critical triangle of sustainable development by enabling farmers to make intertemporal decisions and better face risks. Yet, in rural areas the presence of transaction costs and covariate risks often cause credit markets to fail to provide credit to all at market interest rate, and the poor bear the highest costs of these imperfection. Vietnam's intervention in the rural credit market is embodied by a 'commercial' bank, the Vietnam Bank for Agriculture and Rural Development (VBARD) and a 'policy' bank, the Vietnam Bank for Social Policies (VBSP), lending microloans to poor households at subsidized interest rate.

Using a rich dataset on farmers' credit transactions and access between 2002 and 2007, we describe the credit market, compare contract terms offered by formal, semiformal and informal lenders, examine participation by wealth terciles and estimate a Tobit model predicting interest rates offered by non formal lenders. Data shows that despite offering competitive contract terms, the formal sector remains a secondary source of credit in Yen Chau, particularly for the poor who also face higher interest rates than other borrowers in the informal sector. We then investigate determinants of demand and access to the two formal credit programs using two-stage Probit models with partial observability. Results indicate a selection of borrowers into the VBARD program that results both from a self-selection of borrowers and a selection by the bank. Estimates show no discrimination of the poor in this program. On the contrary, we find the the poor are more often denied credit access by VBSP even though its program is targeted to them. 'Community imperfections' and the fact that credit's agents incentives are too much drawn on the repayment performance of clients are seen as two main factors encouraging mistargeting. We also explore econometrically whether demand for formal and informal credits interact so as to assess the substitutability potential between both sectors, and find weak evidence of an interdependence. Finally, using the propensity score matching approach, we assess quantitatively the impact of the government's micro-credit program on households' livelihoods. Results do not show significant impact. Analyses thus reveal a number of inefficiencies in the credit policy in terms of financial sustainability, poverty outreach and welfare impact that need to be addressed. We derive policy recommendations to address these inefficiencies. Results nevertheless show that the credit market is relatively fluid. Thanks to good levels of trust and of social interaction between individuals, the informal sector is able to provide credit to all at moderate interest rates.

Finally, in the third article, we investigate whether the high degree of ethnic diversity observed in the study area undermines social capital. Social capital has appeared as a powerful concept to capture community features, such as norms and networks that enable members of a society to act collectively and/or overcome market failure. The literature indeed points at ethnic heterogeneity as a factor inhibiting social interactions, and thus hindering collective action. Dissected landscapes in mountainous areas have caused complex settlement patterns overtime resulting in a high cultural and linguistic diversity. In Yen Chau, heterogeneity within villages was further induced by resettlement policies conducted by the government after the independence from France as a way to integrate ethnic minorities in the nation's political project. This particular setting offers a unique occasion to test the exogenous effect of ethnic heterogeneity on social capital outcomes. We develop two econometric models, the first one predicting individual participation and the second analyzing households' investment in social network capital. Results show an ambiguous effect of ethnic heterogeneity on participation. The extent and direction of this effect appear to depend on the type of organization, its political nature and of the type of good that are managed, whether public or club goods. Second, we do not find evidence of a direct effect of heterogeneity on households' level of social network capital measured by its borrowing capacity from friends and relatives. However, we find heterogeneity to have an indirect impact through its interaction with variables capturing households' identity and the participation of their member in local organizations. While the place of birth appears as a significant determinant in homogeneous settings, this effect vanishes as heterogeneity increases. On the contrary, participation of household members to have a positive effect on households' level of social network capital, an effect that becomes significant in heterogeneous communities that increases as communities become more diverse. These results thus do not confirm theoretical predictions of a negative relationship but show instead that heterogeneity can induce a certain dynamism in social relations by favoring the establishment of bridging connections which in turn can foster innovation, cross-cultural learning, ease social mobility, and eventually encourage sustainable development.

The contributions of this thesis are twofold. First, it identifies sources of success and failure in the current institutional framework to promote sustainable development in Vietnam's mountainous areas from which we derive important policy recommendations. Evidence in this thesis highlights limitations of the top-down approach that dominates public intervention in mountainous areas of northern Vietnam. These interventions are costly and often not successful in enhancing equity, efficiency and the environmental sustainability of resource use. This stresses the need for the Vietnamese government to further enhance the functioning of incentive-based mechanisms in the economy. In this perspective, the clarification of the land reform objectives, the development of a land market, the promotion of independent and financially sustainable financial institutions, the reinforcement of the legal system, and the support of the emergence of an independent civil society are all measure that may encourage sustainable development in Vietnam's mountainous regions.

Beyond these policy-oriented findings, this thesis also contributes to a growing body of the literature studying micro-economic impacts of institutions, and delivers in this perspective several general findings. As any economic agents, farmers have anticipations and interpret signals. Therefore, transparency and credibility of the state in implementing a policy are two important features to ensure its success. Second, social capital, thanks to positive externalities notably on financial markets or in the management of natural resource, has strong implications on the economic performance of a society. This should encourage governments to design policies aimed at promoting social interactions particularly among people of diverse cultural and social backgrounds.

List of Abbreviations

ADB	Asian Development Bank
ATE	Average Treatment Effect
ATT	Average Treatment Effect on the Treated
CEM	Committee for Ethnic Minorities
CEMMA	Committee for Ethnic Minorities and Mountainous Areas
CIA	Conditional Independence Assumption
DFG	Deutsche Forschungsgemeinschaft
DV	Dependent Variable
GDP	Gross Domestic Product
GSO	General Statistical Office
FAO	Food and Agriculture Organization of the United Nations
ha	hectare
ha/cap	hectare per capita
HEPR	Hunger Eradication and Poverty Reduction
HH	Household
IV	Instrumental Variable
m.a.s.l.	meter above the sea level
LURC	Long term Use Rights Certificates (also referred to as Red Book)
MFI	Microfinance Institutions
MO	Mass Organization
MOLISA	Ministry of Labor, Invalids and Social Affairs
NGO	Non Governmental Organization
NIE	New Institutional Economics
NMR	Northern Mountainous Region
NVN	Northern Vietnam
OLS	Ordinary Least Square
p.a.	per annum
PCA	Principal Component Analysis
PEN	Poverty-Environment Nexus
PES	Payment for Environmental Services
PPP	Purchasing Power Parity
PPS	Probability Proportional to Size
PSM	Propensity Score Matching
PRA	Participatory Rural Appraisal

SCT	Soil Conservation Technologies
UNDP	United Nation Development Program
VBARD	Vietnam Bank for Agriculture and Rural Development
VBP	Vietnam Bank for the Poor
VBSP	Vietnam Bank for Social Policies
VHLSS	Vietnam Household Living Standard Survey
VND	Vietnamese Dong
WTO	World Trade Organization

The currency in Vietnam is the Vietnamese Dong (VND). The exchange rate in 2007 was of 16 000 VND for 1 US$.

Contents

List of Tables

List of Figures

1. Introduction

1.1 General introduction

Vietnam's economic achievement over the past twenty years constitutes one of the most successful development stories of the last century (Glewwe et al., 2004). Classified among the world's poorest countries at the end of the 1980s, Vietnam is now expecting to join the list of industrialized countries by 2020 (ADB, 2006). After a decade of remarkable success in the 1990s[1], Vietnam has continued to progress in 2000–2010, and is ranked among the fastest growing economies of this decade (with an average annual growth rate of the Gross Domestic Product (GDP) of 7.3% between 2000 and 2012 (World Bank, 2011))[2]. Moreover, economic growth has been pro-poor. The latest estimates from the Vietnam Household Living Standard Survey (VHLSS) indicate a nation-wide poverty incidence in 2008 of 13.1% (share of population living with less than 1.25 US$ in Purchasing Power Parity), a dramatic reduction from 49.7% in 1998 and 63.7% in 1993 (GSO, 2011).

The transformation of institutions has been at the heart of Vietnam's transition strategy, and according to many observers, is a major key to the country's success (Macours and Swinnen, 2002; Cornia and Popov, 2001; Montes, 2001; Rozelle and Swinnen, 2004; Ravallion and van de Walle, 2008b). The *Doi Moi*[3] reform program was enacted in 1986 at the VIth congress of the the communist party to guide the transition from a centrally planned economy to a market-oriented system. Most of the reforms were initially undertaken in the agricultural sector, where in 1981 about 70% of the population was employed. After 28 years of collectivization (1960–1988) the objective of the government was to transform the incentive structure and return small farmers to the center of decision-making. With this perspective, land rights were transferred for most

1 The edited book by Glewwe et al. (2004) provides a good overview and detailed analyses of the results achieved during this decade

2 The Vietnam Development Reports published each year by the World Bank also documents extensively the achieved progress and remaining challenges in Vietnam. The reports of years 2003 and 2006 have focused on poverty reduction and social protection, and those of years 2004 and 2009 have examined institutional advancements.

3 *Doi Moi* is translated in English as renovation.

agricultural land to farmers (since 1988), markets were gradually liberalized, a state-owned agriculture and rural development bank (the VBARD) was created in 1990, and a variety of anti-poverty programs targeted at poor rural households were implemented through the Hunger Eradication and Poverty Reduction program (HEPR, nowadays referred to as National Target Program for Poverty Reduction). The transformations undertaken thus entailed a deep redefinition of the roles of the state, the market and communities in rural areas in regulating resource allocation and organizing rural life.

Achievements in the agricultural sector have been spectacular (Macours and Swinnen, 2002; Rozelle and Swinnen, 2004; Kirk and Tuan, 2009). As a striking example, from a situation of extreme poverty and hunger at the start of the reform, Vietnam became within a few years a net rice exporter (since 1988). The country now ranks among the world's top food exporters (second exporter of rice and coffee in the world (FAO, 2011)).

Yet, this success was not experienced on an equal scale accross the country. Nation-level figures hide important disparities between urban and rural populations, upland and delta regions or between ethnic minorities and the Kinh majority. The Northern Uplands region in which this thesis takes place was the poorest region of the country in 2008. Poverty incidence was still twice as large there as in the rest of the country in 2008, and the pace of poverty reduction has also been slower in the past years compared with the rest of the country, as shown in Figure 1.1[4].

Mountain people in Vietnam and elsewhere in South East Asia regions are among their countries' poorest populations (Blyth et al., 2002; Zeller et al., 2010; Akramov et al., 2010). The reasons for this development gap lie principally in disadvantageous geographic and political conditions. Greater remoteness, lower endowments in arable land, and ecological fragility cause mountain people to have limited access to markets, infrastructure and technologies, and subject them to more economic risks. Those living in the mountains are not only put at a geographic disadvandage, but are politically marginalized. In these regions, the dissected landscape has caused complex settlement patterns overtime, resulting in a rich cultural and linguistic diversity (Blyth et al., 2002, p. 20). Cultural barriers coupled with geographical remoteness cause mountain ethnic minorities to be less politically organized than their urban and lowland counterparts. They are less likely than other groups to influence public choice. As a consequence, policies designed in faraway urban centers by people of different socio-cultural

4 This figure shows expenditure poverty rates using the World bank and GSO poverty line (280 Thousand VND/month/capita in 2008).

Fig. 1.1: Poverty rate in Vietnam and Northern Uplands (1998–2008)

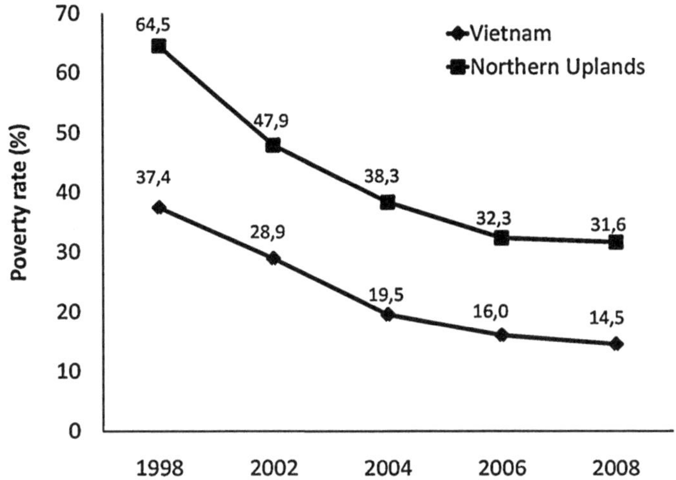

Source: GSO (2011)

backgrounds do not often address adequately their economic, cultural and ecological concerns (Jamieson et al., 1998; Akramov et al., 2010). As stated by Platteau (2000, p. xxii), "*The fact that tribal communities of Asia and Latin America are typically embedded in national entities ruled and dominated by societies with opposite background characteristics is bound to affect their economic and other performances (in terms of wealth, education, etc.) in a different way from what would obtain more homogeneous tribal societies.*"

Yet, as this introduction will show, public intervention is needed to address the complex linkages between environmental, economic and social conditions characterizing mountainous areas in South East Asia (Zeller et al., 2010). Achieving sustainable development in the Upland regions is important, not only from a human and economic perspective, but also because mountain ecosystems provide invaluable services to the rest of the country, which include clean water, food, energy, biodiversity, recreation and protection from environmental disasters (such as floods or landslides) (Blyth et al., 2002; Ahlheim et al., 2009).

The rapid population growth in the uplands of Vietnam over the last 15 years, combined with the intensification and expansion of agricultural systems into fragile areas, has contributed to accentuating the pressure on natural resources. In the northwest region, the population increased by 34% between 1995 and 2010, while it grew only by 20.3% in the rest of the country (18.6% if we exclude the Central Highlands) (GSO, 2011). The scarcity of natural resources has increased the risk that the economy get trapped in a vicious circle whereby poverty and environmental degradation mutually reinforce each other and durably undermine economic development. The existence of strong linkages between livelihoods and environmental protection in the presence of resource scarcity implies that in fragile areas poverty reduction, economic growth, and environmental conservation are complementary goals that must be jointly addressed (Reardon and Vosti, 1995; Duraiappah, 1998; Holden and Binswanger, 1998; Dasgupta et al., 2005). Economic organizations[5], such as the state, markets and local communities and the resulting institutional framework play an essential role in addressing these objectives (Dasgupta and Mäler, 1995; Duraiappah, 1998; Holden and Binswanger, 1998).

This thesis investigates to what extent the institutional framework currently in place in mountainous areas addresses the poverty-environment nexus (PEN), that is, the set of complex linkages between poverty and the environment. Drawing on a conceptual framework that underlines the role of institutions and resulting incentive structure in addressing the nexus, the thesis examines the functioning and impact of three institutional dimensions – land institutions, financial markets, and social capital – that play a critical role in and are particularly relevant to the current context of Vietnam's uplands.

The rest of the introduction is structured as follows: Section 1.2 describes the conceptual framework used in this thesis, and is followed by Section 1.3 which provides background information on Vietnam's land reform and the rural finance policy. Section 1.4 details the specific objectives and reveals the research hypotheses tested research. The outline of the thesis is detailed in Section 1.5.

1.2 Conceptual framework: institutions, poverty and the environment

This section provides the conceptual framework that has guided the elaboration of research questions and the research hypotheses explored in this thesis. The following subsections describe linkages between poverty (equity) and the environment in the agricultural sector (in 1.2.1); define institutions, demonstrate their predominant

5 The distinction between organizations and institutions is often confusing in the literature. According to Hayami (2001, p. 221), an organization is defined as a "*functional body organized by a set of rules*" while institutions are the "*set of rules to organize people into the functional body*".

role in the PEN, and explain why they often fail to address it (in 1.2.2). Finally, the last subsection 1.2.3 details the role of land institutions, financial markets and social capital play in this framework shows how the state, market and communities interplay in their definition.

1.2.1 Linkages between poverty and the environment in agriculture

Small farmers in developing countries derive most of their livelihoods from the exploitation of natural resources such as soils, water and vegetation. While most of these resources are regenerative or renewable, an inappropriate use (or overuse) leads to their degradation and depletion until their exhaustion.

The causal link between environmental degradation and poverty is evident. In a context of growing population and increasing scarcity of natural resource, and when technologies are not available, the depletion of one of farmers' main livelihood resources has a direct negative impact on their wealth. In fragile ecosystems, natural resource degradation often causes and exacerbates the occurrence of natural disasters: soil erosion and deforestation for instance increase the likelihood and accentuate negative consequences of floods and landslides on livelihoods. Added to climatic and market-related risks, these natural disaster reinforce risks and uncertainty in fragile areas. When insurance mechanisms or credit markets are imperfect or missing, these risks and uncertainties threaten livelihoods of vulnerable farmers and undermine their decision making capacity. The potentially negative consequences of water pollution on health constitutes a further threat on livelihoods, not only for local populations but also at a larger scale. Finally, as the availability of essential natural resources shrinks, competition for resources will increase raising risks of conflicts.

Poverty is also pointed as a source of environmental degradation. Farmers make decisions intertemporarilly regarding resource use, depending on the resources available today and those that will be left tomorrow given the impact of today's behavior. Therefore an important element of resource-related decision making is the discount rate, defined as the intertemporal marginal rate of substitution between the farmer's present and future utility (Pender, 1996). Poor farmers lacking financial and physical capital, typically face high discount rates (Pender, 1996; Holden et al., 1998). Their utility of future consumption is only weakly weighted compared to the utility of today's consumption needs, inducing them to (over)exploit resources today without considering the impact on tomorrow's resource availability. Small farmers lacking access to capital also lack investment capacities to diversify their activity away from farming, and are doomed to rely heavily on the exploitation of low access-cost primary resources such as natural resources. Many conservation technologies, require initial investments or incur high opportunity costs in the short term. Farmers lacking investment capacities are unlikely to undertake such investment (Holden and Binswanger, 1998).

However, as pointed by Reardon and Vosti (1995); Duraiappah (1998) and Scherr (2000) focusing on poverty as a main cause of environmental degradation is misleading. Empirical evidence indeed shows that environmental degradation linked to agriculture occurs as well in wealthy environment. The inverted U-shaped environmental Kuznets curve, empirically documented in the 1992 World Bank development report (World Bank, 1991), even predicts environmental degradation to increase along with the level of economic development but to decrease beyond a certain certain level. Scherr (2000) cites case studies showing how small farmers have been able to adapt environmental degradation through cheap and ingenious technologies (for instance, the contour stone bunding in Burkina Faso (Dutilly-Diané et al., 2003)). To others, the extent to which poverty induces environmental degradation appear to depend much on the type of poverty considered – where "type" refers to the asset categories in which households are poor, to the distribution of poverty across households in a society, and to the type of environment problem that takes place (Reardon and Vosti, 1995). In the same line, other authors have distinguish between exogenous and endogenous and by such, argue that extent to which environment and poverty interplay depends in fact on other factors, such as market and other institutional failure. Poverty that results from market and institutional failure affects the environmental sustainability (Duraiappah, 1998). These controversies imply that wealth-enhancing policies while necessary will not be sufficient to address the nexus. Successful interventions will thus address the sources of poverty rather than its consequences, address institutional aspects reinforcing the linkages between poverty and the environment and will often be context-specific.

The important synergies between households' livelihood and environment resources imply that poverty reduction, economic growth and environmental sustainability can no longer be treated as separate objectives but must be jointly addressed in rural development strategies. These three objectives constitute the critical triangle of sustainable development described by Vosti and Reardon (1997).

1.2.2 The role of institutions

Institutions are defined by North (1990, p. 3) as "*the rules of the game or more formally, the humanly devised constraints that shape human interaction*", and refer to the formal (such as laws) and informal rules (such as customs) that regulate human relationships in an economy.

Authors of the New Institutional Economics (NIE) perspective argue that the economic importance of institutions and of organizations lie on the presence of important transaction costs in the economy (Williamson, 1979; Bardhan, 1989). Imperfect information has focused much attention. Because of the occurrence of opportunistic behaviors and bounded rationality in the economy, information asymmetries are source of important transaction costs which impede the functioning of markets (Akerlof, 1970). Transaction costs arise from the need to screen and to monitor transaction partners and from costs needed to enforce property rights. This framework has been used to explain the emergence and persistence of important institutional failure in rural areas of developing countries, such as credit rationing (Stiglitz and Weiss, 1981) or sharecropping (Stiglitz, 1989), and in general to explain why many markets fail[6] or are even missing in these areas.

Market and institutional failures incur important costs for the economy. When property rights are incompletely specified or imperfectly enforced, markets for environmental resources do not emerge. As a consequence, the negative externalities related to the misuse of natural resources are not accounted for creating no incentives for their user to protect them (Dasgupta and Mäler, 1995). de Janvry et al. (1991) show moreover how market failure undermine production efficiency, but also the reactivity of farmers to economic policies – such as subsidization – and thus contribute to increase the inefficiency of public intervention in the agricultural sector.

1.2.3 The role of land institutions, financial markets and communities

Institutional and market failures are the source of important inefficiencies in the economy and tend to reinforce the PEN in the agricultural sector. In this thesis

6 Markets are said to fail when they induce an allocation of resources which that is suboptimal in the sense of Pareto, that is when the welfare of some could be improved without deteriorating the welfare of others.

we focus on three institutional dimensions: land institutions, rural finance institutions, and social capital. The following paragraphs explain below why these institutions are critical, why they often fail in rural contexts, and how these failures can be addressed.

Land institutions

It is widely acknowledged that well-defined land property rights encourage the natural resource management, yield positive efficiency effects and have the potential to reduce poverty (Meinzen-Dick and Knox, 2001; Deininger and Feder, 2001).

First, well-defined and enforceable land rights, by ensuring the right-holder to reap the future benefits of today's investments, and letting him bear the consequences of his mismanagements, create incentives for investment and natural resource management. Second, well-defined property rights allow land markets to develop and enable the transfer of land from the less productive to the more productive farmers, generating important efficiency gains in the rural economy. Finally, land constitutes with labor the main resource small farmers can use to ensure their livelihoods. Well-defined and secure land rights also enhance farmers' livelihood, by securing small farmers' access to land resources, enabling them to sell this resource at market price, or to use it as a collateral in credit transactions.

As the competition for resources increases, defining and enforcing property rights becomes more costly. As explained by Platteau (2000, chapter 3 and 4), these costs can be so prohibitive that they have prevented the natural emergence of efficient land institutions in many developing countries. Public intervention is needed to assume these important costs. Land registration and titling policies have appeared in this perspective as the most efficient intervention and have been promoted likewise in many developing countries. The issuance of titles offer land users an incontestable mean to claim and enforce their use rights on a defined land area, and thereby increase tenure security, and enable land transactions. Land titles enable small farmers to use land as collateral in formal banks (Feder and Akihiko, 1999; Deininger and Feder, 2001; Deininger, 2003)[7].

---

7 There are nowadays important discussions in the development sphere about land titling policies, which are on the one hand very costly, and on the other hand not always successful in creating tenure security, particularly in Africa areas where traditional land tenure systems are very complex. On this question, see Atwood (1990), Bromley (2009) and Meinzen-Dick and Mwangi (2009).

Rural financial markets

Many decisions in farming activities require intertemporal decision making, i.e. decisions that are made in the present and entail future outcomes. In ecologically fragile and populated areas, the availability of future resources strongly depends on today's actions. Moreover risk and uncertainties are predominant features of economic lives. The functioning of financial markets (which encompass markets for saving, insurance and credit services) – sometimes referred to as 'intertemporal markets' – enabling farmers to make intertemporal decisions has strong implications for the PEN.

The access to financial services enable farmers to bear risks and to smooth consumptions over seasons and, as such, has direct positive effects on welfare (Zeller and Sharma, 2000). Second, a better access to financial services, credit in particular, reduces opportunity costs of capital, and thereby encourage technical progress and the use of labor-saving technologies (Diagne et al., 2000). Finally, well-functioning markets are likely to encourage environmentally sustainable practices, by enabling farmers to make intertemporal investments and reducing their discount rates[8].

In rural areas of developing countries, information asymmetries, the covariance of income and saving behaviors, and the high level of uncertainty cause financial markets to fail. The risks of loan default, which arise with moral hazard and adverse selection, incur for the lender important screening, monitoring and enforcement costs. These costs induce informal lenders to raise interest rates up to prohibitive levels, or to ration poor borrowers based on their perceived creditworthiness (Stiglitz and Weiss, 1981). Formal banks, as a way to screen borrowers and reduce default risks, require the deposit of collateral as guarantee causing small farmers lacking adequate collateral to be excluded.

The negative consequences caused by failures on financial markets, in terms of equity, efficiency and environmental sustainability require external intervention. Models of interventions have varied through time and across countries. Repressive interventions (e.g. through interest rates ceilings) and subsidization which dominated interventions up to the 1980s have been mostly unsuccessful in addressing market failure and rural poverty (Conning and Udry, 2005). This lead to the emergence of a new paradigm advocating for institutional innovations enabling lenders

---

8 The relation between poverty, credit access and discount rates has been studied and put in evidence by Pender (1996) in India, and by (Holden et al., 1998) in Ethiopia and Zambia.

to overcome information asymmetry and enforcement problems (Zeller, 2003). The success of the microfinance 'revolution' demonstrated that through institutional innovations, the access of poor households to financial services can be enhanced, while ensuring financial sustainability of institutions. The role of government in this paradigm is subject to different interpretations. While some see the financial sector mostly as a private sector, others highlight the public good character of financial innovations and insist on the necessity for the state to support emerging institutions in this sector (Lapenu, 2000).

The role of communities and social capital

As seen above, information asymmetries, and the lack of enforcement mechanisms are two major factors causing institutional and market failures. Public intervention is then required to enhance the functionning of the economy, for instance by defining formal rules that can reduce transaction costs, or, by substituting the market in the allocation of public goods. Yet, the government may not always succeed to address market failure, in particular when imperfect information and enforcement problems are the main factors causing markets to fail. Remote public agents may not do better than the local traders (or lenders) in accessing information or enforcing rules.

Communities, in comparison, have a comparative advantage both in information access and in their capacity to enforce rule through social sanctions. For these reasons, they provide according to Hayami (2001, p. 286), "*a principle of organization critically needed to correct the failures of the market and the state, and, thereby, to support modern economic development.*" This observation have led many development organizations and government to allocate an increasing role to communities in the design and implementation of development projects – particularly in the field of natural resource management projects – through the so-called community-driven development projects.

Yet, considering that communities are all uniform in their ability to enforce rules and to share information is misleading. Social capital, defined by Putnam (1995, p. 664) as the "*features of social life – networks, norms and trust – that enable participants to act together more effectively to pursue shared objectives*" has appeared in this perspective as a powerful concept to capture the features that enable community members to exchange information, enforce rules and to act collectively. As explained later in this thesis (section 6.2) social capital remains difficult to define and to capture as a as a single tangible concept. It is best seen as a multifaceted concept, that encompass both cognitive and structural aspects, and has both a community and an individual outcomes (such as social networks) (Woolcock and Narayan, 2000).

Theoretical work and empirical evidence suggest that the level of social capital in a given community is – at least partly – endogenously determined. Individuals decide whether to join a group, to cooperate, based on the economic, social or institutional conditions they face (Woolcock and Narayan, 2000). Community feature, such as income inequality, ethnic diversity, geography, the pre-existence of clear cooperation rules are all potentially important factors to explain why people cooperate better or form social networks faster in some areas than others. Empirical research on the formation of social capital remains, up to date, very scarce. Theoretical evidence however suggests ethnicity and identity to play an important role on social capital formation. In particular, ethnic heterogeneity is seen as a factor inhibiting social interactions due to linguistic and cultural barrier, social sanction effects, and preference mechanisms[9].

To sum up, this section showed that institutions define the incentive structure in the economy and, as such, play a critical role in the poverty-environment nexus in mountainous areas, by inducing small farmers to choose (or not) sustainable livelihood strategies. Drawing on the sustainable livelihood conceptual framework (Scoones, 1998), and based on theories detailed above, Figure 1.2 summarizes the conceptual framework of this thesis and presents its outline.

1.3 Background information

This section briefly describes the land reform and rural credit policy of Vietnam's government. Both policies are described in more details within the chapters of this thesis, in sections 3.2 for the land reform, and 5.1 for the credit policy.

1.3.1 The land reform

Vietnam's land reform has induced a quasi-privatization of land access. In this system, the land officially belongs to the Vietnamese people and is managed by the state on its behalf, land users are transferred income and control rights for a delimited time period. The reform began officially in 1988, year at which the resolution 10 enacted the end of collective farming, and transferred use and control rights to

9 A review of the literature on this issue is provided in chapter 6, section 6.2.

Fig. 1.2: The poverty-environment nexus and structure of the thesis

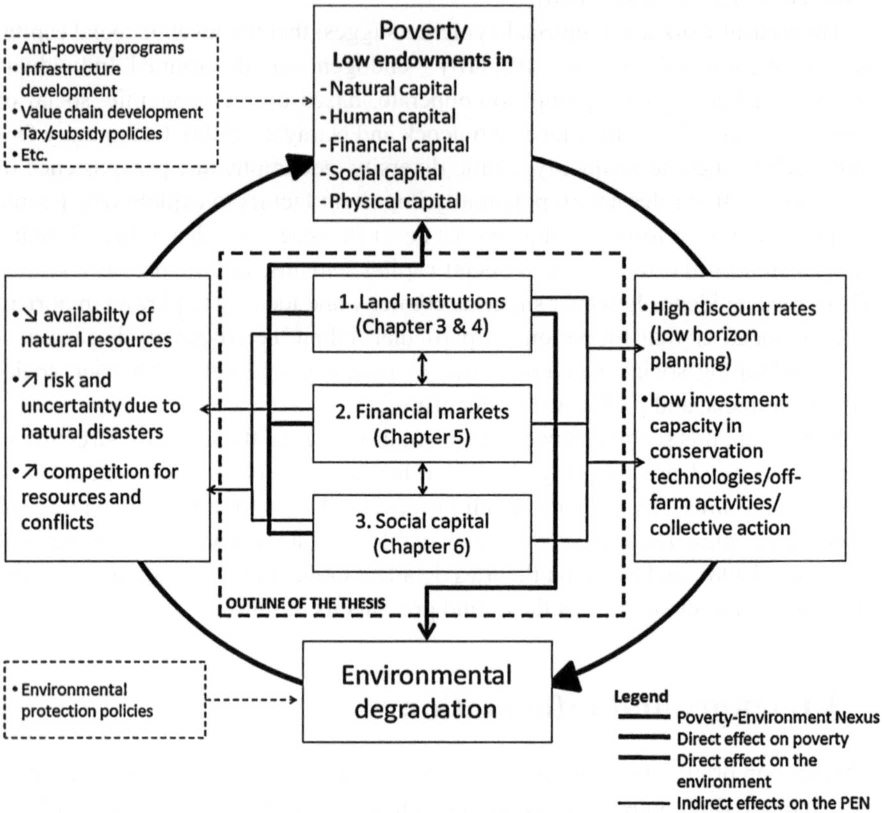

Source: Own figure

individual farm households. The 1993 land law granted users with five rights: the right to exchange, transfer, mortgage, inherit, and lease out the land. Land users received land use right certificates (LURC, sometimes referred to as Red Book), which entitle them to claim and use these rights for a period varying between 20 years for annual crop land to 50 years for perennial crop land, and to an unlimited time period for residential land. In the law, the state, through local authorities, maintained heavy control over transactions, rental contracts, and more importantly over general land use plans. A new law was voted in 2003 to complement the previous one, and was aimed at facilitating land transactions, extending the rights of both household spouses (by registering titles under both spouses names), and including the right to sublease land to the previous bundle.

This policy represents one of the largest land titling programs ever implemented throughout the developing world with around 11 million titles issued in 2000 (Do and Iyer, 2008) and was assessed by observers and researchers as egalitarian (Ravallion and van de Walle, 2004, 2008a). In mountainous regions, the enforcement of the land law has been contested by some ethnic groups who returned instead to traditional land management systems when collective farming ended (Corlin, 2004; Sikor, 2004). Uncertainty persists on whether these rights will be reallocated at the end of the use right term.

1.3.2 The rural credit policy

Vietnam's rural credit policy is embodied by two state-owned banks (World Bank, 2008). The Vietnamese Bank for Agriculture and Rural Development (VBARD) was created in 1990 and acts as a profit-oriented commercial bank supporting the development of rural areas by providing loans to agricultural and non-agricultural enterprises. VBARD's interest rates remain however partly controlled by the state (Duong and Izumida, 2002; Dufhues, 2007).

The other bank is the Vietnamese Bank for Social Policies (VBSP), former Vietnam Bank for the Poor (VBP). Its mandate is to offer microcredits at preferential interest rates to a targeted population, mainly the poor. The bank transfers its lending activities to village-based mass organizations (MO), and is a hybrid institutions between village fund and a microfinance institution (MFI). The bank does not collect savings and its program remains heavily subsidized by the government.

As highlighted by Dufhues (2007), both banks constitute in most provinces the sole formal credit supply available to farmers. Unfair competition conditions due to high levels subsidies in the VBSP, and political constraints have prevented other organizations to establish. As a consequence, rural areas have not yet seen the development of a competitive microfinance sector that is independent from state intervention.

1.4 Specific objectives and research hypotheses

The economic transition in Vietnam has entailed, in rural areas, a deep redefinition of the role of the state, the market and the communities in allocating resources and organizing rural life. As seen above, upland areas in Vietnam are characterized by dissected landscapes, low infrastructure development, high poverty rates and ethnic diversity, features that have altered both the implementation of the reforms as well as their outcomes.

Based on the conceptual framework detailed above, this thesis examines the performance of the current institutional framework in Vietnam's mountainous region in addressing the poverty-environment nexus. Its overall objective is to identify sources of successes and failures, so as to derive policy recommendations likely to support sustainable development in the region.

While the political and institutional economy of sustainable development in Vietnam's mountain is a broad and complex issue due to the complex linkages, this thesis concentrates on selected research issues that have appeared to the author, during field work and in view of the literature particularly relevant. As such this thesis offers a partial but substantial insight into this question.

This research, as already mentioned is organized around three research areas investigating the impact of Vietnam's land reform, the functioning of the credit market and the impact of the state policy, and the formation of social capital within communities. The specific questions addressed in each the four chapters following this introductions are listed below:

Research question 1 – Impact of the land titling policy.

a. Has the land titling policy induced more tenure security?
b. Does the reform encourage the adoption of soil conservation practices?
c. Did the reform lead to the emergence of a land market?
d. What is the distributional outcome of the reform and, if any, of the land market?

Research hypothesis 1 – Following the conceptual framework in section 1.2.3, the issuance of land titles, by offering land users an incontestable mean to claim and enforce their use rights on a defined land area is expected to increase tenure security, and thereby to create incentives for conservation investments. Moreover, by granting land users with transfer rights, it is expected that the policy creates necessary conditions for a land market to establish. Following the literature on Vietnam's land policy, it is hypothesized that the policy has been implemented in an egalitarian way. However, the distributional impact of the land market is ambiguous as it is conditioned by other factors, such as credit access.

Research question 2 – Functioning of the credit market and impact of the state policy.

a. How well does the credit market functions? who participates? and what are the market shares of the different sectors?

b. Does the State intervention induce a more equitable and efficient credit allocation than other sectors?
c. Can the state-governed formal sector substitute the informal sector?
d. Is the government micro-credit program yielding positive impacts on household welfare?

Research hypothesis 2 – Following the conceptual framework in section 1.2.3, it is hypothesized that the dissected landscape, in addition to the ethno-linguistic diversity, and the covariant risks are the source of important transaction costs and information asymmetries, and cause credit market to fail. Thus the intervention of the government through a commercial and a policy bank, by expanding formal credit supply, and by providing micro-loans at low interest rates to poor farmers is expected to yield positive impacts on farmers' welfare.

Research question 3 – The formation of Social capital.
a. Does ethnic heterogeneity undermine social interactions and the participation in local organizations?
b. What is the impact of ethnic heterogeneity on the formation of social networks?

Research hypothesis 3 – In view of the theoretical literature on the role of identity and ethnic heterogeneity on the formation of social capital, it is hypothesized that the high level of ethno-linguistic diversity in Vietnam's mountainous areas inhibit social interactions and the formation of social networks.

Beyond the primary policy-oriented objective, a second and not less important objective is to contribute to the growing body of literature studying microeconomic impacts of rural institutions. There is now a large consensus in the economic literature that institutions greatly matter for development. Yet, according to Pande and Udry (2006), the recent empirical literature establishing a link between well-defined institutions and economic performance has mostly focused on demonstrating a positive effect on economic growth through cross-country data analyses. The microeconomic literature establishing causal links between institutions and economic behavior on the other hand remains mostly theoretical. As stated by these authors, there is a need to expand our knowledge and understanding of how institutions work at the microeconomic level, and to assess mechanisms at place. This evidence will help to identify the policy options and institutional arrangements likely to enhance the functioning of the economy in an equitable and efficient manner. The institutional transformation that have taken place in Vietnam during the transition and its implication in the PEN offers in this perspective an interesting case study.

1.5 Outline of the thesis

This thesis is organized around seven chapters. After this introduction, chapter 2 presents the study area, and the quantitative and qualitative data collection. Chapter 3 presents an article addressing questions 1a and 1b, e. g. on the impact of the land reform. This chapter is completed by chapter 4 which investigates empirically research questions 1c and 1d, relative to the establishment of a land market, and the distributional impact of the reform. The research question 2, relative to the functioning of the land market is addressed in chapter 5, while the last research research area relative to the formation of social capital is investigated in chapter 6. Finally, in chapter 7, we discuss the results and derive policy recommendations that are potentially relevant to enhance sustainable development in Vietnam's mountainous regions.

2. Presentation of the study area and data collection

The research conducted in this thesis takes place in Yen Chau, a rural district located in the North Western region, Son La province. The area is located 300 km west of Hanoi, along the National 6 road linking Hanoi with the West of the country (see map in figure 2.1). The district shares typical agroecological and socio-economic characteristics with other mountainous areas in South East Asia and for this reason it has been chosen as a focus research area by the *Uplands Program*, SFB 564 the collaborative research program on sustainable land use systems in South East Asia under which this research was conducted. The population grew by 2.4% annually between 1988 and 2006 according to the district' statistical department. This is twice as high as recorded in the rest of the country.

A household survey was conducted between 2007 and 2008 in Yen Chau, within the subproject F2.3 of the Uplands Program. Two project leaders, and three PhD students (including the author) were part of the project. The sampling procedure used to select households is described extensively different sections of the thesis (section 3.4, 5.2 and 6.3.2)[10], therefore it is not being detailed here. The sample is composed of three hundred households, equally spread in twenty villages (fifteen households randomly selected in each village), randomly selected among all farm households inhabiting the district[11]. A map of Yen Chau depicting the location of the sample villages is shown in figure 2.1. Five ethnic groups are represented in our sample, the Thai (75%), the H'Mong (15%), and the Kinh (9%). Other groups (the Sinh Mun an Kho Mu) represent only a small fraction of the district's population (less than 1%). Table 2.1 describes the twenty sample villages, their size, ethnic composition, elevation and economic classification.

Household- and village-level questionnaires were used to gather data on a wide range of issue, including among other detailed modules on household food and non food expenditure, credit access and transactions, land access and agricultural practices, social capital, etc.[12] The questionnaires were developed together with

10 The three papers presented in this thesis use the same survey data.

11 Except for villages and households located in the four communes bordering Laos for which research permits were difficult to obtain.

12 This survey covered three PhD topics including this one which explain the variety of topics discussed.

Fig. 2.1: Yen Chau district and sample villages

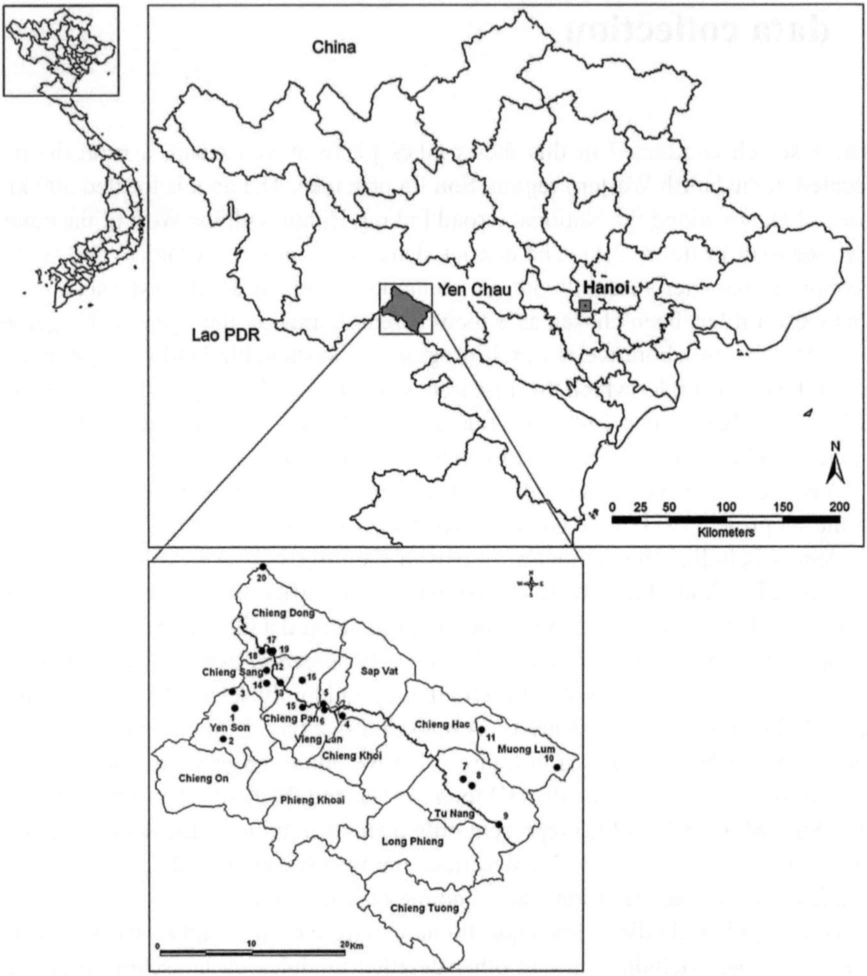

other members of the research team in the section Rural Development Theory and Policy of the Institute of Agricultural Economics and Social Sciences in the Tropics and Subtropics of the University of Hohenheim. Both questionnaires contain most of the data that have been used for this research.

The questionnaires were administered within four rounds: the first two were implemented in March-May 2007, the third one in July-August 2008 and the last one in December 2007–January 2008. A team of interviewers composed mostly

Table 2.1: Description of sample villages

Commune	Map	Village	Pop.[a]	Main Ethnicity[b]	Elevation (m.a.s.l.)	Zone[c]	Class.[d]
Yen Son	1	Cho Long	676	Thai	943	II	A
	2	Ban Dan	657	Thai (70%), Kinh (30%)	682	II	A
	3	Chieng Hung	542	Kinh	673	II	A
Sap Vat	4	Ban Dong	260	Thai	311	I	A
Vieng Lan	5	Na Va	246	Thai	295	I	A
	6	Kho Vang	424	Thai	286	I	R
Tu Nang	7	Na Ten	343	Thai	470	II	A
	8	Na Khoang	395	Thai	440	II	P
	9	Ta Lang Thap	631	Thai	303	II	R
Muong Lum	10	Ban Dao	247	H'mong	964	III	P
	11	Khau Khoang	315	H'mong	821	III	P
Chieng Sang	12	Mai Ngap	568	Thai (53%), R Kinh (47%)	367	II	R
	13	Ban Dan	1030	Thai	349	II	R
	14	Chieng Sang	892	Thai	369	II	A
Chieng Pan	15	Na Xanh	228	Thai	315	II	R
	16	To Pang	470	Thai	355	II	A
Chieng Dong	17	Luong Me	860	Thai	410	I	A
	18	Then Luong	505	Thai	448	I	R
	19	Dong Tau	930	Thai	404	I	A
	20	Keo Bo C	142	H'mong	995	I	P

[a] Number inhabitants in 2006/2007.

[b] Here we indicate the main ethnicity, and do not indicate other ethnic group when those constitutes only a small fraction (<30%) of the village population. As explained in chapter 6 however, many villages are in fact mixed.

[c] Communes are officially classified in three economic zones according to infrastructure development and poverty rates. I indicates the richest zone, and III the poorest. The poor communes are targeted in some infrastructure and poverty reduction program such as program 135.

[d] Subjective classification by village heads of their village within their commune. P=Poorer, A=Average and R=Richer.

of local citizens (native or living in the district) and of students from Hanoi University of Agriculture was hired for this purpose. Most of the local interviewers were familiar with the local language spoken in the area (Thai) and a large share of interviews could be conducted in this language. In remote H'mong villages, the resort to local translators was sometimes needed. Most interviewers had experience working as agricultural extension officers prior to work for the project, and helped to adjust the questionnaire to local conditions. Careful pre-tests were also conducted. Enumerators were trained extensively prior to the start of the survey to ensure that misinterpretation or misunderstanding would not affect the quality of data collected. Cross-check questions were included in the different questionnaires to verify accuracy of information provided by respondents. Once filled, questionnaires were checked on site by a team leader (usually one of three PhD researchers involved in that project), and rechecked by a data entry operator. All questionnaires were entered twice (using SPSS Data entry double entry mode) to avoid typing errors. Finally, the data were carefully cleaned by the PhD students using for some part of the data (e.g. consumption data) routines previously elaborated in similar projects. All these precautions were undertaken to maximize the quality of the data.

The method of inference in this thesis is mostly quantitative. Research hypotheses are tested by means of econometric models, that are designed and estimated for this purpose. The objective is to provide quantitative evidence on research questions enumerated in section 1.4, and by such, to identify causal, systematic and quantifiable relationships.

The advantage of quantitative approaches is that they enable a large-scale analysis and generate results that are representative in a given area. Limitations are nevertheless important. Beyond the risk of measurement errors – which still remain despite careful checks – many information and relationships are difficult to capture quantitatively and through structured interviews conducted by a third person. This problem is particularly acute in studies that cover social and institutional dimensions. To analyze the impact of an institutional arrangement, a sound understanding their definition and functioning is required. While information on formal rules such as law texts may be easy to access (because kept in a written form), informal rules exist mostly in unspoken and unwritten form.

Moreover, institutions in place (or the *de facto* rules) are often a combination of informal and formal rules. It is indeed common that the *de jure* formal rules are interpreted locally and adjusted to local norms, and eventually differ from the observed *de facto* institutions. The sole investigation of the legal literature provides only partial and incomplete information, and a sound exploration on-site is needed to complement these sources. Likewise, a strong divergence between *de jure* and *de facto* rules was observed in Yen Chau, particularly in land institutions.

Hence, in addition to quantitative data, focus group discussions were conducted within each sample villages, with a group of elderly and/or knowledgeable members to trace back the land allocation and settlement history in each village. In addition, informal interviews were conducted with key informants. Land officers at the district and commune level were interviewed several times, the local branches of the VBARD and VBSP were visited, and their directors interviewed. Discussions with the district, commune and village leaders of the women, farmer and other mass organizations were conducted. This was complemented by household-level informal interviews conducted in single households that were identified either as special or representative cases, regarding credit access or land use. In some cases, semi-structured questionnaire was complemented with the use of participatory rural appraisal (PRA) tools. Likewise, visual timelines were drawn during the focus group discussions, and also during some household interviews to collect chronological data.

Although qualitative data analysis is not performed in this research, the qualitative information collected has influenced much of its outcomes. First, this information helped in the formulation of research questions and in designing the questionnaire, particularly on land issues. It was also used to triangulate research findings, and as complement the quantitative analyses in each chapters of this thesis.

3. Land titling policy and soil conservation practices in the northern Uplands of Vietnam[13]

Abstract

In Vietnam, a quasi-private property regime has been established in 1993 with the issuance of exchangeable and mortgageable long-term land use right certificates. Using primary qualitative and quantitative data collected in a mountainous district of Northern Vietnam, this paper investigates the role of the land policy in the adoption of soil conservation technologies by farmers. This issue is of crucial importance in the region where population growth and growing market demands have induced farmers to intensify agricultural production. While poverty has been reduced, environmental problems such as soil erosion, landslides, and declining soil fertility have become more severe over the past years. Our findings suggest that despite farmers' awareness of erosion, soil conservation technologies are perceived as being economically unattractive; therefore, most upland farmers continue to practice the prevailing erosion-prone cultivation system. Focusing on agroforestry as one major soil conservation option, we estimate household and plot level econometric models to empirically assess the determinants of adoption. We find that the possession of a formal land title positively influences adoption, but that the threat of land re-allocations in villages discourages adoption by creating uncertainty and tenure insecurity. The analyses reveal that these two effects interact with each other but are of small magnitude. We conclude that the issuance of land titles is a necessary but not sufficient prerequisite to encouraging the adoption of soil conservation practices. However, current practices remain economically unattractive to farmers. This deficiency needs to be addressed by interdisciplinary research and complemented by strong efforts by local authorities to promote sustainable land use.

3.1 Introduction

Beginning with the Doi Moi policies in 1986, Vietnam has engaged in important institutional reforms in order to lead its economy from a centrally planned to a market oriented system. In rural areas, reforms were designed to strengthen

13 This chapter has been published as Saint-Macary, C., Keil, A., Zeller, M., Heidhues, F., Dung, P., T., M., (2010) Land titling policy and soil conservation practices in the northern Uplands of Vietnam, *Land Use Policy*, 27(4), 617–627 - The Appendix section is not part of the published article.

farmer's decisionmaking capacity as a way to boost production and encourage the protection of natural resources. In this perspective, the land allocation policy and the issuance of long-term land use right certificates to households have been among the most important measures taken by the state.

Our research focused focuses on the Northern Mountain Region (NMR), an area inhabited predominantly by ethnic minorities[14]. Rapid population growth over the past 20 years has led to an increasing scarcity of arable land driving agricultural production up onto the steep slopes. Moreover, the increasing demand for food and feed from urban areas has induced farmers to further intensify the production of maize, an erosion-prone crop, on the hillsides. Soil erosion and landslides have become important issues as they have lead to reduced soil fertility in the uplands, sedimentation of lowland water reservoirs, irrigation channels and paddy fields as well as severely damaged road infrastructure. Unchecked, these practices could affect the greater population as they threaten the sustainability of agricultural production in the mountainous areas as well as local infrastructure and, at a larger scale, the quality of drinking water in the lowlands. Soil conservation technologies (SCT) are thus a major tool for sustaining livelihoods and agricultural production at the individual farm level and ensuring food security for coming generations.

The determinants of soil degradation and the adoption of conservation practices have been widely investigated in the literature. While techniques to limit erosion on cropped fields are numerous and diverse, most of them share two common features: first, they incur opportunity costs in terms of land and labour that are not available for cropping activities any longer; these costs may be rather high, especially if SCT compete with profitable cash crops. And, second, SCT enhance soil fertility in the medium to long run only (Lutz et al., 1994). Hence, the adoption decision of SCT involves intra- and intertemporal utility tradeoffs (McConnell, 1983; Grepperud, 1997). Poor farmers living in an environment of incomplete credit markets, insecure land tenure, and short planning horizons are unlikely to invest in such technologies, notably because of potential risks for present or future consumption (Pender, 1996; Shively, 2001).

Thus, state interventions that secure land tenure, provide safety nets to rural households, and relax constraints on liquidity via credits to farmers are expected to reduce poverty and encourage natural resource conservation and protection

---

14 The Vietnamese ethnic, the Kinh, represent about 82% of the country's population. The rest of the population is composed of 53 other ethnic groups located for most of them in mountainous regions.

(Lutz et al., 1994). In particular, it is expected that by securing land tenure, improving access to formal credit, and increasing land transactions, land titling policies increase investment incentives and foster rural development (Deininger and Feder, 2001). The formalization of property rights is, according de de Soto (2000), a major step towards development and poverty alleviation as it enables poor households to transform their assets into productive capital. Notwithstanding this consensus, empirical studies conducted in different countries do not find that property rights have large and positive effects as investment incentives including formal land titles in particular (see Bromley, 2009 for a review). Rather, it seems that the issuance of formal land titles has not necessarily led to increased tenure security and to higher investments on land. Legal pluralism and the co-existence of formal and informal land institutions have been identified as one cause of inefficiency of land institutions, causing in addition high enforcement costs (Platteau, 1996; Meinzen-Dick and Pradhan, 2003). Absent or imperfect labour or credit markets may as well limit investment capacities and the efficient allocation of land among farmers, making land institutions inefficient. Recent empirical studies have found that land titles may favor certain types of households more than others, for example the richest households in Paraguay (Carter and Olinto, 2003) or the powerful ones in Ghana (Goldstein and Udry, 2008).

Regarding soil conservation practices, Soule et al. (2000) find that land owners in the United States have a higher propensity to adopt such technologies than cash or share-renters. Shively (2001) also concludes that land tenure is a significant determinant of SCT adoption in the Philippines. Lutz et al. (1994) and Holden et al. (1998) both conclude that policies aimed at securing land tenure are necessary but not sufficient prerequisites to encouraging the adoption of soil conservation practices.

In Vietnam, using national living standard survey data, Do and Iyer (2008) find that the 1993 land law has significantly increased the allocation of land to perennial crops, although the effect is only of small magnitude. In contrast to most of the existing literature on Vietnam's land titling policy, this paper presents a small-scale and detailed study which allows to control for local factors potentially relevant for explaining the impact of land titles on investment incentives.

The objectives of the paper are (1) to describe how the land titling policy has been implemented in the NMR, (2) to assess upland farmers' knowledge and adoption of soil conservation technologies, and (3) to identify the influencing factors of the adoption of such measures by household and plot level econometric models and controlling for knowledge. We focus in particular on the role of the land titling policy on fostering such investments.

After outlining the land reform and its implementation in Vietnam in Section 3.2, we describe the research area in Section 3.3 and present our data and sampling procedure in Section 3.4. The estimation strategy for our regression models is described in Section 3.5, results are presented in Section 3.6. In Section 3.7 we discuss the results and draw conclusions.

3.2 Land reform and implementation in the NMR

Prior to 1981, agricultural land, means of production, and production output were fully managed by the collectives, namely the village cooperatives. The first reform of tenurial contracts was implemented by Directive 100 in 1981, which allowed farmers to keep the surplus produced in excess of the contracted output for home consumption. The series of directives that followed were designed to increase the flexibility of the land management and taxation systems in order to raise investment incentives, increase production levels, and overcome food insecurity (Que, 2005). The collective farming system officially ended with the 1988 land law (Resolution 10) and the allocation of collective land to private households according to family size. The land is still owned by the Vietnamese people (and managed by the state); farmers initially received land use rights for a period of fifteen years.

Pursuing the decollectivization effort further, the 1993 land law granted five rights to land users: the right to exchange, transfer, mortgage, inherit, and lease out the land. Land use certificates (known as Red Books) were issued to users for a period of 20 years for annual crops and of 50 years for perennial crops. The law confirms that the land is publicly owned, it is therefore a quasi-private land management system. Through local authorities the state maintains control over transactions, rental contracts, and more importantly over general land use plans.

This policy represents one of the largest land titling programs ever implemented throughout the developing world with around 11 million titles issued in 2000 (Do and Iyer, 2008) and was assessed by observers and researchers as egalitarian (Ravallion and van de Walle, 2004, 2008a). Its implementation however has been a costly process, and has not been evenly achieved in the regions. In 1998 less than half of the total area in NMR had been allocated with a certificate (Do and Iyer, 2003). The considerable administrative costs involved for measuring land, issuing and registering the certificates and the lack of qualified personnel were the main reasons for this slowness (Haque and

Montesi, 1996). Secondly, in mountainous regions, the enforcement of the land law has been contested by some ethnic groups who returned instead to traditional land management systems when collective farming ended (see Corlin (2004) on H'mong villages; Mellac (2006) and Sikor (2004) on Thai communities). Sikor (2004) reports cases of conflicts in Thai communities in Yen Chau district where village heads refused at first to allocate paddy fields as directed by the Red Book, preferring instead the traditional allocation system in which the village head regularly reallocates paddy fields to villagers. Corlin (2004) reports the same types of conflicts over rights allocation and the conflict with ancestral rights in H'mong communities. In some regions, as in Yen Chau, these conflicts have been solved by reallocations accompanied by strict enforcement through the provincial government (cf. Section 3.6.1). Uncertainty persists on whether these rights will be reallocated at the end of the use right term (20 years). Despite repeated interventions of the ministry of agriculture stipulating that no reallocation will occur, an important share of farmers still expects the contrary and is reluctant to engage in land transactions[15].

3.3 The research area

The study region is Yen Chau district, a mountainous area inhabited primarily by ethnic minorities (85%). The largest ethnic groups are the Thai (Black Thai) with about 55% of the district's population, followed by the H'mong who account for 20%, and the Kinh, "ethnic Vietnamese" who represent 13% of the district's population. The Thai and the Kinh were the first settlers in the area and occupied the lowlands, while later arrivals, such as the H'mong, settled mainly in the highlands (Neef et al., 2002). Lowland villages along the highway benefit from greater access to infrastructure (i.e. markets, paved roads, and irrigation systems) and are relatively better off than villages located in higher altitudes. Between 1988 and 2006 the districts' population rose by 50% at a growth rate of 2.4% annually.

Farmers nowadays cultivate two main crops: rice as a subsistence crop grown in irrigated paddy fields of the lowlands and maize as a cash crop in the uplands. The rapid development of the livestock sector accompanied by rising maize

15 cf. Interview of Mr. Dang Hung Vo, Vice Minister of Natural Resources and Environment (28/08/2007) [ONLINE (in Vietnamese)]: http://www.agro.gov.vn/news/newsDetail.asp?targetID=2128, [last accessed, 27/11/2008].

prices and the shortage of rainfall have pushed farmers to intensify maize production and abandon rice cultivation in the uplands. Thus, the area allocated to maize cultivation has more than tripled over the last twenty years while the area allocated to upland rice has decreased by 27%, according to the district statistical office. In 2007, on average 86% of farmers' upland area was cropped with maize from April to September and left uncovered the rest of the year, exposed to wind and rainfall. As a consequence, the area is susceptible to erosion leading to an increasing incidence of sedimentation and landslides during the rainy seasons over the past years.

3.4 Sampling and data collection

A household survey was conducted in 2007/2008 in Yen Chau district. In selecting the households, a cluster sampling procedure was followed where a village-level sampling frame was constructed encompassing all villages of the district[16], including information on the number of resident households. First, 20 villages were randomly selected using the Probability Proportionate to Size (PPS) method (Carletto and Morris, 1999). Next, 15 households were randomly selected in each of these villages using updated village-level household lists. Since the PPS method accounts for differences in the number of resident households between villages in the first stage, this sampling procedure results in a self-weighing sample (Carletto and Morris, 1999). In total the database consists of 300 households and 2279 agricultural plots, of which 2059 are operated by farmers and 1190 are upland plots, i.e., rain-fed plots dedicated to crop production other than paddy rice. Both the household and plot samples are representative at the district and village levels. The survey covered a wide range of topics, including questions on quantitative and qualitative features and on the effect and perceptions of soil conservation technologies.

In addition, we conducted focus group discussions using semi-structured interview guidelines to collect information on village history and composition and to reconstruct the chronology of the land allocation process. A visual timeline was elaborated with respondents in order to facilitate these recall questions, and a standardized questionnaire gathered the necessary background data. The respondents were part of the current and former village boards.

---

16 Except for villages in four communes bordering Laos, for which research permits were difficult to obtain.

3.5 Estimation strategy

3.5.1 The household-level model

We estimate the determinants of farmers' decision to invest in agroforesty[17], and in particular the effect of the land titling policy on adoption incentives. Since aggregating different technologies that imply different costs and benefits over time may be misleading, we focus on agroforestry which is one of the most widely known measures against soil erosion in the study area and is also perceived to be one of the most effective (cf. Section 3.6.2).

The investigation of adoption determinants in a population where the diffusion of innovation is incomplete may lead to biased estimates (Diagne and Demont, 2007). Selection bias arises when exposed and unexposed farmers differ in their propensity to adopt the technology[18]. This may be the case for at least two reasons. First, knowledge acquisition is part of the farmers' adoption decision and therefore endogenous and, second, for efficiency reasons agricultural extension may especially target farmers or communities with a high innovative capacity. In Section 3.6.2 we show that knowledge diffusion for agroforestry is incomplete in the study area suggesting the use of a selection bias correction model. Our problem can be written as follows:

$$y_{1i} = \begin{cases} 1[\beta X_1 + \upsilon > 0] & \text{if} \quad y_2 = 1 \\ 0 & \text{otherwise} \end{cases} \tag{3.1}$$

$$y_{2i} = 1[\gamma X_{2i} + \upsilon_i > 0] \qquad \forall_i \in [1, N] \tag{3.2}$$

where N is the total population; y_{1i} and y_{2i} are binary dependent variables indicating the adoption and knowledge status of the ith household respectively; X_{1i} and X_{2i} are

17 "Agroforestry is a collective name for land-use systems in which woody perennials are deliberately grown on the same piece of land as agricultural crops and/or animals" (Lundgren, 1982). By agroforestry, we refer to a cultivation technique consisting in planting trees and/or shrubs on a cultivated land as a way to limit soil erosion and improve soil fertility. The plants mostly used in the study area are wild tamarind (*leucaena leucocephola*), teak trees and pine trees.

18 In this context selection bias is also termed exposure bias in the literature, e.g. by Diagne and Demont (2007).

vectors of regressors; u_i and υ_i are the error terms, we assume that they are jointly bivariate normally distributed. The covariance matrix is:

$$\operatorname{cov}(u_i, \upsilon_i) = \begin{pmatrix} \sigma_u^2 & \rho \\ \rho & \sigma_\upsilon^2 \end{pmatrix}$$

We use a Heckman full maximum likelihood procedure to jointly estimate the probability of knowing and adopting the technology and control for selection bias (Heckman, 1979). The model predicts household's probability to adopt and maintain agroforestry on at least one of its plots conditional on variables X_{1i} and on knowing agroforestry[19] (Table 3.1).

Table 3.1 summarizes the explanatory variables (contained in X_{1i} and X_{2i}). Following literature on knowledge acquisition and learning (Feder and Slade, 1984; Foster and Rosenzweig, 1995; Conley and Udry, 2001), we expect that information access is closely linked to education, social capital, the possession of communication assets, access to the agricultural extension service, and income. The social capital variable measures how well the household is connected with mass organizations[20] in its village by assessing how easily help is obtained if necessary, which is referred to as vertical Demont (2007) connections in the following. In addition, we include a variable on households' participation in the farmer union as a measure of horizontal social capital. The daily per-capita expenditure variable is used as a proxy for wealth[21]. Other variables control for access to agricultural extension, human capital, and possession of communication assets.

Among the regressors of adoption, control variables account for major household characteristics (the number of active members, education, age of the household head, and wealth level), soil, and farm characteristics, as well as geographic location. In addition, we include a regressor indicating whether material support was received by the household to implement agroforestry. Material support includes labour, in-kind inputs (seeds, seedlings or fertilizer, for instance) or cash support.

---

19 The probability $P(y_{1i} = 1 | y_{2i} = 1, X_{1i})$ is derived in Wooldridge (2002a, pp. 477–78 and 570–71).

20 In Vietnam, the mass organizations play a crucial role and are present at all administrative levels (from the village to the state). They are composed of six unions representing women, farmers, veteran, elderly, youth and the fatherland front union. In addition to participating in major village decisions, these organizations carry out multiple tasks: from extension agents to rural bank staffs.

21 The survey contained an extensive module on households' food and non-food expenditures, asset values, and remittances received, from which this variable is calculated.

Table 3.1: Description and summary statistics of household-level variables

Variable	Description	Mean	S.D.
Knows agroforestry+	HH knows agroforestry as SCT	0.42	0.50
Age*	Age of the household head	43.16	12.66
Education*+	Highest educated member has high school certificate	0.05	0.23
Actives*	Number of actives (non disable, aged between 18–60)	2.54	1.26
Vertical connections	Number of problems for which it is easy to get help from Union[a]	2.01	2.00
Radio+	HH possesses a radio	0.18	0.38
Extension service	Subjective score on access to extension service (1 = lowest, 5 = highest)	3.1	1.06
Farmer union+	HH participates in the farmer union	0.79	0.41
Expenditure*	Daily expenditure per capita in Thousand VND[b]	15.40	6.48
Elevation*	Elevation of the house in meter above the sea level (m.a.s.l.)	520.33	241.87
Adoption+	HH has adopted and uses agroforestry on at least one plot	0.12	0.32
Poor soil	Share of area with poor soil (%)	30.89	33.61
Medium soil	Share of area with medium soil (%)	56.11	36.57
Relative upland size+	Area operated per capita > village average	0.61	0.49
Support+	HH received support to implement agroforestry	0.07	0.26
Credit constraint+	HH is credit constrained in the formal sector	0.27	0.45
Titled land	Share of titled area on total area operated (%)	70.64	39.07
HH exp. real.+	HH experienced at least one reallocation	0.09	0.28
Villagers exp. real.	Share of villagers who experienced reallocation[c] (%)	8.32	10.51
HH expects real.+	HH believes reallocation is likely to occur	0.79	0.40
Villagers expecting real.	Share of household in village expecting reallocation[c] (%)	79.01	13.54
			292 Obs.

* these variables are present both in the main and the selection equations.

\+ indicate dummy variables.

[a] Respondents were asked to assess the easiness in receiving help from village mass organization and village head to (i) borrow money for education; (ii) borrow money for health expenses; (iii) borrow money for any positive event; (iv) borrow money for any negative event; (v) borrow a water buffaloe; (vi) ask for labour.

[b] Variable logged in the regression.

[c] Excludes the household himself.

Such support has been provided either by governmental or non-governmental organizations to encourage certain farmers to adopt agroforestry. Hereby, several goals may have been pursued, and different types of households targeted: firstly, the focus may have been on farmers with a low investment capacity; secondly, extension organizations may have targeted influential or exemplary farmers as a way to disseminate a technology (using a demonstration plot); finally, it may also have been intended to enforce adoption in areas of strategic importance, such as easily visible locations close to the main road[22]. However, statistical tests show no systematic differences between supported and unsupported households regarding potential influencing factors of adoption (human, social, and financial capital). We therefore conclude that the attribution of support was random regarding characteristics that also influence adoption and that, consequently, we do not face an endogeneity problem.

We hypothesize that improved access to credit is conducive to the adoption of agroforestry based SCT because it relaxes liquidity and/or consumption constraints, and reduces farmers discount rates. This will lead to a higher value being attached to benefits from reduced soil erosion that accrue in the future Pender (1996); Holden et al. (1998). We use a binary variable indicating whether a household is credit constrained on the formal credit market. Following Zeller (1994) we consider farmers to be credit constrained if they did not apply for credit for fear of rejection or if they applied for a loan but were partially or fully rejected by the lenders[23].

The effect of the land titling policy on adoption incentives is captured by five variables. A first variable measures households' share of upland area operated under a land title. A positive and significant coefficient would indicate that land title is perceived as a guarantee of tenure security thus encouraging farmers to engage in soil conservation. Apart from being registered and operated under a title, the land can be (*i*) leased from private households for a defined period of time and/or a fixed payment; (*ii*) lent or given by private households with no payment and for an undefined time period; (*iii*) leased or borrowed from the village fund land; or (*iv*) cultivated without agreement or informally purchased (Table 3.2). With this variable on land title, we test our main hypothesis that tenure is perceived to be

22 It was mentioned that some farmers having their plot located close to the National road had been strongly encouraged to implement hedgerows on their field so as to create positive impressions on officials and visitors passing by the area.

23 The literature on credit and technology adoption suggests that this variable is endogenous (i.e. correlated with unobserved factors of adoption such as entrepreneurial capacity). The test of endogeneity conducted on this variable does not reject exogeneity, with various specifications and various measures of credit access. We therefore treat this variable as exogenous in the model.

Table 3.2: Tenurial arrangement and year of land acquisition

	% Plots[a]	**% Household**[b]	**Year acquired (mean)**
Operated under title	74.9	81.5	1989
Operated without title	25.1	50.5	2002
- Leased in from private household	6.8	19.5	2006
- Borrowed to private household	10.1	19.9	2002
- Village land fund	4.8	13.0	2000
- No agreement	2.3	8.2	1997
- Non registered purchase	1.0	2.1	2002

[a] Upland plots operated in 2007 (T=1190).
[b] Households using at least one plot under such arrangement (N=292).

more secure when land is operated under a title than under any of the other tenurial arrangements. Several empirical studies have found evidence that tenure security may be endogenous to investment, as farmers may undertake certain investments to secure tenure and obtain land titles (Besley, 1995; Place and Swallow, 2000; Brasselle et al., 2002). In our case, the fact that land titles have been distributed to all households at a certain point in time (cf. Section 3.6.1 and Table 3.2) excludes this risk of endogeneity, but the correlation of the land title variable with unobserved factors of adoption is might be a source of endogeneity. We conduct some test, presented in the Appendix of this chapter[24].

As outlined in Section 3.6.1, the implementation of the land policy in the study area has resulted in successive reallocations, and a majority of farmers expect further reallocations to take place before the end of the use right term. While the issuance of a land title was supposed to empower farmers as decision makers over the use of their land, the successive reallocations may have sent the contradictory signal that the state remains the primary decision maker over land issues. We include four variables to capture these effects: (*i*) a dummy variable indicating whether the household has experienced a reallocation on its upland plots, (*ii*) the share of households in the village (excluding the respondent household) that have experienced upland reallocations, (*iii*) a dummy variable indicating whether the household believes that a reallocation is likely to occur before the end of the use

24 The Appendix Section is not part of the published articles, because of space constraints.

right term, and (*iv*) the share of villagers (excluding the respondent household) expecting such a reallocation. These variables are only very weakly correlated[25] giving no cause for concern regarding multicollinearity. We include both household and village variables since decisions regarding land use are partly made at the village level. As a post-socialist country, Vietnam has a long tradition of collective decision making. This is particularly true in the rural areas, especially regarding land use. Villages in the study area are mostly homogenous in ethnicity and even often constituted by households from the same clan. Consequently, the ties among villagers are usually strong, and the social life within a village is very intense, providing good information circulation within a village. Historically and nowadays, the 'Vietnamese village' is considered a strong entity in itself. Both the colonial administration and the communist party have later tried to instrumentalize the villages in order to impose political projects (see Bergeret (2003, p. 30–33) for a detailed historical review of the "Vietnamese village myth"). After the decollectivisation these structures have remained or even been reinforced. Sikor (2004) and Wirth et al. (2004) show for Yen Chau how the organisation of villages has challenged the implementation of the land law. At a larger scale, Kerkvliet (1995, 2005) shows how 'everyday politics' within Vietnamese villages have contributed to transform the national policy at the time of the decollectivisation. Based on this evidence, we believe that it is plausible that villagers' experiences and opinions significantly matter in a household's decision regarding the adoption of agroforestry.

3.5.2 The plot-level model

Apart from investigating which factors influence a household's decision to adopt agroforestry, we are also interested to know where within a farm agroforestry is adopted. Indeed, as shown in Table 3.2, a significant share of households in our sample operates both titled and untitled land (households cultivate on average four upland plots). A household-level model is unable to capture the effects of tenure, soil characteristics, and other plot-specific variables on adoption. Hence, we also estimate a plot-level model for the adoption decision. This approach has been widely applied in the literature to estimate the impact of land tenure on adoption incentives (Besley, 1995; Hayes et al., 1997; Pender and Fafchamps, 2001; Hagos and Holden, 2006).

In the plot-level model we are not able to correct for exposure bias, as the selection regarding knowledge occurs at the household level. However, the test

25 The highest correlation among these four variables (0.2) is found between the household and village-level variables regarding experiences of reallocations.

of independence of equations (Wald test) applied to the household-level model does not indicate any selection bias (cf. Section 3.6.3). Therefore, we run the plot-level model based on the households knowing the technology only. The plot-level model can be written as follows:

$$y_{3ij} = 1\left[\alpha X_{3ij} + \beta X_{1i} + \varepsilon_{ij} > 0\right], \quad \forall j \in \left[1, T_i\right] \quad j \in \left[1, N^k\right], \quad N^k < \mathrm{N} \quad (3.3)$$

where y_{3ij} indicates the adoption status of agroforestry by farmer i on plot j, T is the number of upland plots operated by household i and the sub-sample of households knowing agroforestry as a SCT. We cluster the standard error at the household level in order to account for heteroskedasticity and for non-independence of observations within a household (Wooldridge, 2006). The plot-specific variables X_{3ij} included in the model are described in Table 3.3.

3.6 Results and discussion

3.6.1 Land allocation in Yen Chau district

Due to the region's formerly low population density and ethnic diversity, land management in the NMR has proceeded differently from the rest of the country. During the period of cooperative farming informal permission was given to individuals to cultivate the uplands bordering the common land as long as collective goals were met.

Table 3.3: Description and summary statistics of Plot-level variables

		Full sample (N)		**Restricted sample (N^k)**	
Variable	**Description**	**Mean**	**S.D.**	**Mean**	**S.D.**
Adopt+	Agroforestry is adopted	0.04	(0.19)	0.08	(0.27)
Poor soil+	Soil is of poor quality	0.30	(0.46)	0.30	(0.46)
Medium soil+	Soil is of medium quality	0.55	(0.50)	0.54	(0.50)
Area share	Area divided by HH farm size	0.23	(0.20)	0.21	(0.19)
Steepness+	The slope is very steep[a]	0.37	(0.48)	0.38	(0.48)
Land title+	Operated under land title	0.75	(0.43)	0.80	(0.40)
		1190 Obs.		567 Obs.	

+ indicates dummy variables (yes=1, no=0)

[a] The slope was assessed by respondents on a scale from one (=level) to five, using a graph for illustration.

With the passing of the first land law in 1988 farmers were encouraged to clear forest and upland areas as only the common land was distributed following official criteria. The application of the 1993 law and the issuance of land use right certificates (LURC) followed a long process: some of the households received a title in 1991 while others, few kilometers away and usually on higher altitudes, received a first title only in 1999.

We summarize the issuance of process in the study area in three phases:

- From 1991 to 1996 : a first allocation was achieved in part of the lowland villages. The allocation was carried out by commune and village officials, with some supervision by provincial-level staff. Respondents reported that despite the official criteria established (family size, soil quality), land was generally allocated to the actual users, without being properly measured. In some Thai villages only upland areas were registered on the certificate and the allocation of paddy land remained under the authority of the village heads.
- After 1998, the provincial administration implemented a second wave of land allocation. Upland villages, where allocation had not yet taken place, were issued the first certificates. In other villages land was reallocated and land titles reissued. During this second wave, provincial officials were actively involved to enforce the law. Agricultural and residential plots, including paddy fields, were allocated following the official criteria, formally measured, and the lands were titled and recorded in cadastral maps. A share of non-allocated land (paddy and upland) was kept as a village fund and controlled by the village board for allocation to newly established households.
- Since 2003 a third allocation wave has been taking place. In villages where reallocation occurred (five out of our twenty sample villages), land titles had yet to be reissued in 2007. The official rationale for this reallocation was firstly to provide land to newly established and landless households and secondly to combine scattered small plots as a way to increase farm productivity, following the land use plan established by the Province[26].

Despite the existence of long-term use right certificates originally issued for twenty to fifty years, some households have seen their land reallocated two times at five year intervals. Whereas only a minority of farmers have been directly affected by reallocations of either paddy fields or upland plots (35% of our sample), 65% live

26 However, in light with the current situation, it appears that this reallocation is linked to a larger scale plan to introduce rubber plantations in the district.

in villages in which a reallocation has occurred. As a consequence, 80% of the sample farmers believe that a reallocation will take place in their village before the end of their 20-year use right, indicating the low level of trust households currently have in land institutions.

3.6.2 Knowledge and adoption of soil conservation technologies

We report in Table 4 descriptive qualitative and quantitative results on farmers' knowledge on soil conservation technologies and adoption behaviour. Farmers were first asked to enumerate the methods they know to limit erosion. Most farmers three quarters know at least one SCT and are therefore aware of problems related to erosion. When looking separately at different technologies, the diffusion rates[27] vary widely. Knowledge on terraces, contour ploughing, or ditch[28] techniques has been spread mostly through social networks, whereas other technologies have been diffused by more formal communication channels such as media and external organizations. With the exception of the hedgerow technique, the governmental agricultural extension service and non-governmental organizations (NGOs) appear as secondary sources of information on SCT.

Table 3.4 reports adoption rates, defined as the share of household knowing a technology and currently using the technology on at least one of their plot. An effectiveness score based on adopters' perception is also reported. It appears clearly that the methods requiring a relatively high input of labor or take up a considerable portion of land (terraces, vegetative contour strips, agroforestry and cover crops) are adopted the least although they are found to be effective. Short-term and low extra-input technologies (contour ploughing or ditches) are more attractive to farmers but are deemed to be less effective. Among adoption constraints (i.e. the main reason given by respondents for not adopting a known technique), the lack of land is frequently cited in the case of vegetative strips, cover crop, or agroforestry. Lack of labour was identified by farmers as an important constraint for building terraces and planting cover crops. Respondents emphasized lacking access to seedlings as a major reason for not adopting agroforestry and, with regard to

27 The term diffusion refers to knowledge in this paper.

28 The ditch technique consists of channels oriented diagonally to the slope of the land so that rain water is captured and channeled off the field. This technique is used for soil conservation rather than water conservation as the channels are rarely connected to the paddy fields.

Table 3.4: Knowledge about and adoption of SCT

	Knowing SCT (N^k) (% of N)[a]	**Knowledge source**[b]	**Currently using SCT (% of N^k)**	**Perceived effectiveness**[c]	**Adoption constraints**[d]
Ditches or channel	56.2	Relative/Neighbor – Own initiative	61.0	5.7	No effective – Labour
Agroforestry	42.5	Extension – Relative/Neighbor – Media	27.4	6.7	Seedlings – Land
Terrace	20.9	Relative/Neighbor – Other area – Media	9.8	7.0	Labour – Too expensive
Contour ploughing	20.2	Relative/Neighbor – Own initiative	88.1	6.1	No erosion – Equipment
Cover crop	12.7	Media – Extension	10.8	7.3	Land – Labour
Vegetative strips	5.8	Media – Own initiative	11.8	6.0	Land – Labour
Mulching	3.4	Media – Relative/Neighbor	20.0	5.7	Labour
Other SCT	5.1	Own initiative – Relative/Neighbor – NGO	66.7	5.8	
TOTAL (at least one)	**74.7**		**53.4**[e]		

[a] N=292: non farmers and farmers growing paddy rice only are excluded
[b] Ranked by frequency of answers.
[c] Users only – Subjective score, from 0 (no effect) to 10 (very effective).
[d] Sample of farmers knowing but not using only. Ranked by frequency of answers.
[e] Share of total households using at least one SCT.

Fig. 3.1: Land reforms and SCT adoption rates (plot level)

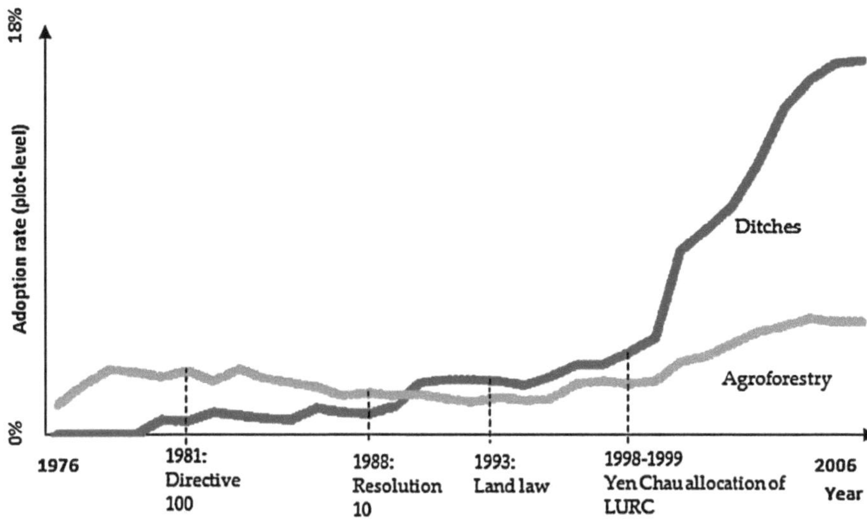

ditches, their ineffectiveness against erosion. The differentiated answers given by respondents show that farmers' perception of costs and benefits over time differ significantly between SCT, so does their adoption decision.

The dynamics of the adoption of selected SCT over time (at the plot level) are shown in Figure 3.1. In 1990, agroforestry was practiced on 1.8% of all the plots operated at that period and ditches were used on 2%. Note that the graph does not account for a cohort effect due to the under-representation of older generations. Nonetheless, one can observe that changes in adoption rates follow institutional changes. The adoption of ditches has clearly accelerated following the first land law in 1988 (decollectivization), while agroforestry has slowed down. The issuance of Red Books in the study area has been clearly followed by a considerable acceleration of adoption of the two technologies, although the increase in the case of agroforestry is less pronounced.

3.6.3 Determinants of knowledge and adoption of soil conservation – the household-level model

Table 3.5 presents results of the household-level probit model with sample selection. In the last column we report the marginal effects on the probability to adopt agroforestry (column (2)). The model has a good predictive quality with 82% of adopters correctly predicted. The upper part of Table 3.5 shows that vertical connections, the possession of a radio, access to the extension service, participation in the farmer union, and the wealth level are positive and statistically significant determinants of farmers' knowledge of agroforestry.

We find that the decision to adopt agroforestry is positively influenced by education and income levels. The farms' soil characteristics, and the size of farmers' upland area relative to the village average are also significant factors. The model predicts that, ceteris paribus, material support received increases the probability of practicing agroforestry by 67 percentage points, while access to formal credit is insignificant. In combination, these two findings indicate a low initial motivation of farmers to undertake such an investment on their own[29], but this interpretation must be nuanced by the fact that the income level is found to be a significant factor.

The share of titled land positively influences households' adoption decision. An increase of this share by one percentage point increases the probability of adopting agroforestry by 0.2 percentage points. Whether the farmer or his neighbour experienced a reallocation of upland plots does not appear to influence adoption, neither does the household's own expectation regarding such a reallocation. The villagers expectation, however, is found to negatively affect farmers' propensity to adopt. An increase by one percentage point of the share of villagers believing that a reallocation is likely to occur reduces the adoption probability by 0.6 percentage points. This is an indication that, because of the tight social organisation of the villages, the general opinion prevails over individual expectations regarding this decision. More importantly, it shows that the reallocation threats, as perceived by the villagers, may discourage the adoption of agroforestry.

---

29 An illustrative example is given by a qualitative case study conducted in one of the H'mong sample villages by social scientists: there, farmers were given seeds and a lump sum of 300,000 VND (around 20 US$) to test hedgerows on their upland fields. In spite of this initial support, only a small fraction of these farmers still maintain these hedgerows nowadays; and if they do so this mostly happens out of fear to be punished otherwise by the administrative office which supported them. Farmers mentioned the lack of profitability of this technique and the competition with their primary cash crop, maize, for land, sunlight, and nutrients as major disadvantages. The findings of this case study are however not fully representative of the situation in the whole of Yen Chau district as indicated by statistical tests: a Hausman specification test concludes that the full sample and the restricted sample (excluding this village) estimates differ systematically. A likelihood ratio test concludes that the restricted sample model fits the data better.

Table 3.5: Household-level model of adoption of agroforestry, probit with sample selection estimates

	Coefficient estimates $P(y_{2i}=1)$ and $P(y_{1i}=1 \mid y_{2i}=1)$ (1)		Marginal effects $P(y_{1i}=1)$ (2)	
	Coeff.	**z-stat[a]**	**dy/dx (x100)**	**z-stat[a]**
y_{2i}: Household knows agroforestry as a SCT (yes=1)				
Age household head	–0.007	(1.22)	–	
Education+	0.550	(1.61)	–	
Actives	–0.041	(0.67)	–	
Vertical connections	0.083**	(2.18)	–	
Radio+	0.641***	(3.18)	–	
Extension service	0.160**	(2.22)	–	
Farmer union+	0.540***	(2.64)	–	
Expenditure per capita (log)	0.346***	(1.97)	–	
Elevation	–0.0004	(1.23)	–	
Constant	–1.695***	(2.57)	–	
y_{1i}: Household uses agroforestry on at least one plot (yes=1)				
Age household head	–0.010	(0.81)	–0.320	(0.85)
Education+	1.306**	(2.39)	48.420***	(2.73)
Actives	–0.106	(0.77)	–3.355	(0.77)
Expenditure per capita (log)	0.740*	(1.74)	23.430**	(2.33)
Poor soil (share)	0.030**	(2.48)	0.945***	(3.26)
Medium soil (share)	0.027**	(2.46)	0.852***	(3.15)
Relative upland size+	0.562**	(2.01)	16.939*	(1.91)
Support+	1.956***	(3.95)	66.619***	(5.05)
Credit constraint+	0.411	(1.42)	13.727	(1.46)
Titled land (share)	0.008*	(1.86)	0.261**	(2.10)
HH experienced reall.+	–0.349	(0.97)	–9.887	(1.10)
Villagers experienced reall.	0.026	(1.55)	0.836	(1.54)
HH expects reall.+	–0.212	(0.60)	–6.711	(0.59)
Villagers expecting reall.	–0.019*	(1.85)	–0.610**	(2.04)
Elevation	–0.001	(1.42)	–0.038	(1.45)
Constant	–3.490*	(1.74)	–	
Observations				292
Censored				168
Log likelihood				–224.3
Estimated ρ (P-value of Wald-test independence equation (ρ=0))				2.45 (0.12)
Correctly predicted (%, cut-off 0.5)				77.3
Adopters correctly predicted (%, cut-off 0.5)				82.3
Adopters correctly predicted (%, cut-off 0.25)				94.1

[a] Robust z-statistics in parentheses: *, (**), [***] significant at 10%, (5%) and [1%] level of error probability.
+indicate dummy variables.

In order to further investigate how this perceived reallocation threat may interfere with the titling policy, we include in a second step the same variable interacted with the variable on titled land[30]. Results are shown in model (3) in Table 3.6. The interacted term is insignificant. The coefficient of the land title variable becomes insignificant, but the coefficient on the villagers' expectation remains negative and significant and more than doubles in magnitude. In accordance with Ai and Norton (2003) we interpret these two coefficient as the effect of the variable when the other variable is equal to zero. On the one hand, this indicates that in absence of a reallocation threat the land title does not influence the adoption decision. But, on the other hand, the impact of a perceived reallocation threat is stronger when the household does not hold a land title. The impact is ambiguous when both variables are different from zero.

In model (4) in the last column of Table 3.6 we introduce the household expectation variable interacted with the land title variable. None of the three coefficients is significant, but the coefficient of the village-level variable remains unchanged.

3.6.4 Determinants of adoption of soil conservation – the plot-level model

Results from the probit estimates of the plot-level model are shown in Table 3.7. The predictive power of the model is rather limited, which can be explained both by the presence of numerous household-level regressors and the low plot-level adoption rate. However, apart from two coefficients (on wealth level and relative upland size), the results are very similar to the ones produced by the household-level model, which indicates that the results are robust.

As in the household-level model, we find soil characteristics to be very important determinants of farmers' decision. Agroforestry is used on poor and medium soils rather than on fertile ones. Farmers' propensity to establish agroforestry on a plot of poor soil fertility is 19 percentage points higher than on a fertile plot. In conformance with the qualitative results, the relative size of the plots also influences the adoption decision at the plot level. We find that 62% of adopters chose the first or the second largest of their plots to implement the technology. Regarding agroforestry, space is a constraining factor as both trees and hedgerows occupy part of the plot, lead to shading of the crop, and compete for nutrients.

30 We thank an anonymous reviewer for this suggestion.

Table 3.6: Household-level model of adoption of agroforestry, probit with sample selection estimates – Interaction tests

	Interaction 1 (3)		Interaction 2 (4)	
	dy/dx x100	**z-stat[a]**	**dy/dx x100**	**z-stat[a]**
y_{3ij}: Household uses agroforestry on at least one plot (yes=1)				
Age household head	–0.285	(0.77)	–0.32	(0.85)
Education+	45.355***	(2.51)	47.364***	(2.66)
Actives	–2.954	(0.68)	–3.194	(0.73)
Expenditure per capita (log)	23.934**	(2.46)	23.53**	(2.39)
Poor soil (share)	0.967***	(3.38)	0.953***	(3.32)
Medium soil (share)	0.879***	(3.30)	0.857***	(3.18)
Relative upland size+	17.256*	(1.94)	16.705*	(1.90)
Support+	67.599***	(5.06)	67.387***	(5.04)
Credit constraint+	14.195	(1.51)	13.909	(1.48)
Titled land (share)	–48.408	(0.74)	14.901	(0.68)
HH experienced reallocation+	–9.712	(1.08)	–9.969	(1.13)
Villagers experienced realloc. (share)	0.830	(1.53)	0.811	(1.48)
HH expects reallocation+	–5.006	(0.44)	–19.418	(0.89)
Villagers expecting reallocation (share)	–1.417*	(1.94)	–0.581*	(1.87)
Elevation	–0.039	(1.48)	–0.040	(1.51)
Villagers expecting realloc. x Titled land	0.009	(1.10)	–	
HH expects realloc. x Titled land	–		0.148	(0.55)
Observation	292		292	
Censored	168		168	
Log likelihood	–224.0		–224.3	
Estimated ρ (P-value of Wald-test)	2.54 (0.11)		2.45 (0.12)	
Correctly predicted (%, cut-off 0.5)	76.7		77	
Adopters correctly predicted (%, cut-off 0.5)	82.3		79.4	
Adopters correctly predicted (%, cut-off 0.25)	94.1		94.1	

[a] Robust z-statistics in parentheses: *, (**), [***] significant at 10%, (5%) and [1%] level of error probability.
+ indicates dummy variables.

Table 3.7: Plot-level model of adoption of agroforestry, probit estimates and interaction tests

	Marginal effects (1)		Marginal effects Interaction 1 (2)		Marginal effects Interaction 2 (3)	
	dF/dx (x 100)	z-stat[a]	dF/dx (x 100)	z-stat[a]	dF/dx (x 100)	z-stat[a]
Age household head	–0.677	(1.30)	–0.064	(1.33)	–0.068	(1.38)
Education (dummy)	7.847*	(1.87)	7.101*	(1.78)	7.619*	(1.86)
Actives	0.424	(0.71)	0.405	(0.70)	0.416	(0.72)
Expenditure per capita (log)	0.791	(0.63)	0.735	(0.62)	0.747	(0.62)
Poor soil+	18.808***	(4.30)	18.796***	(4.28)	19.272***	(4.29)
Medium soil+	8.339***	(3.57)	7.835***	(3.54)	8.000***	(3.55)
Area share	8.91***	(4.06)	8.521***	(4.04)	8.636***	(4.08)
Steepness+	1.503	(1.47)	1.461	(1.49)	1.486	(1.52)
Relative upland size+	1.601	(1.38)	1.537	(1.39)	1.598	(1.45)
Support+	14.812***	(4.91)	14.432***	(4.93)	14.614***	(4.93)
Credit constraint+	0.714	(0.52)	0.708	(0.53)	0.792	(0.59)
Land title+	2.621***	(2.62)	–1.093	(0.14)	1.029	(0.46)
HH expects realloc.+	–2.043	(1.18)	–1.857	(1.11)	–10.277	(1.56)
Villagers expecting realloc.	–0.136***	(3.03)	–0.189**	(2.16)	–0.128***	(2.95)
Elevation	–0.004	(1.42)	–0.004	(1.43)	–0.004	(1.46)
Vill. exp. real. x Title	–		6.489	(0.73)	–	
HH exp. real. x Title	–		–		2.691	(1.08)
Observations	567		567		567	
Log likelihood	–105.2		–105.0		–104.8	
Pseudo R-squared	0.33		0.33		0.33	
Correctly predicted (%, cut-off 0.5)	92.9		92.6		92.7	
Adopters correctly predicted (%, cut-off 0.5)	24.4		22.2		22.2	
Adopters correctly predicted (%, cut-off 0.25)	49.0		55.6		55.6	

[a] z-statistics in parentheses are based on robust standard errors clustered by household.
*, (**), [***] significant at 10%, (5%) and [1%] level of error probability.
+ indicates dummy variables.

As already observed in the household-level model, support is found to be an important determinant of the adoption decision (although the measured effect is of smaller magnitude) while access to credit is insignificant. According to the plot-level model, receiving outside support increases the probability of establishing agroforestry by 5 to 24 percentage points (95% confidence interval).

Finally, variables capturing land policy have a significant effect. In this model we omit the variables regarding experiences of land reallocations as their coefficients were found to be insignificant, and their inclusion reduced the predictive quality of the model[31]. We find that plots operated under a land title are more likely to be covered by agroforestry than other plots. The marginal effect, significantly different from zero at the 1% level, ranges from 0.7 to 4.6 percentage points (95% confidence interval). Similarly to the household-level model, we find that the household's personal expectation does not influence adoption at the plot level, but that the same variable measured at the village level does have a significant negative impact ($P < 0.01$), the marginal effect amounting to a 0.13 percentage-point decrease in the adoption probability for a one percentage-point increase in the share of villagers expecting a reallocation.

In models (2) and (3) we include again the two variables on expected reallocation interacted with the land title dummy variables. The results are in line with the conclusions drawn from the household model. Note further that in model (3) the variable on household's expectation becomes significant with a negative coefficient at the 12% level.

3.7 Conclusion

The reforms of land institutions by the state in the 1990s were intended to increase tenure security, establish a land market, thereby increase investment incentives and boost agricultural production while fostering natural resource conservation. Empirical evidence shows that overall the reform has been a success in this respect. However, the implementation of the reforms has been a long and complicated process due to the intention of the authorities to satisfy both equity and efficiency objectives. In the mountainous regions, the enforcement of the law has been partly opposed by ethnic minority communities. Moreover, we find that in Yen Chau the state maintains substantial control over land resources by periodically carrying out land reallocations, thus sending contradictory signals to farmers and contributing to a perception of tenure insecurity despite the existence of land use right certificates.

31 Results of the models including these variables are available upon request.

We developed household and plot-level econometric models to investigate the role of the land policy in farmers' decision to invest in soil conservation technologies, in particular agroforestry. Confirming previous empirical findings we find that tenure security does influence farmers' decisions regarding such investments. Firstly, the presence of a land title positively influences adoption both at the household and plot level. Secondly, the reallocation threat perceived at the village level is a discouraging factor for the adoption of soil conservation practices. This effect is even stronger when land is operated without a title. However, the positive land title effect disappears when no reallocation threat is perceived, indicating that the latter factor is a substantial one in explaining farmers' behaviour. These effects remain small in magnitude, however. Land tenure policy therefore appears as a necessary but limited tool to foster environmental protection. In our study area, a better clarification of the land use objectives may help to secure land tenure and to promote a more sustainable land use than the environmentally damaging practices that currently prevail.

Using qualitative and quantitative data, this paper investigated also other determinants of investments in soil conservation practices. Although the majority of farmers are aware of soil erosion and know methods to mitigate the problem, adoption rates of these methods remain low. Farmers perceive these techniques to be economically unattractive as they compete with the main cropping activities for scarce land and labour resources. In the case of agroforestry, we find that adoption is influenced by education and the wealth level of the households, but more strongly by attributes of farmers' land, such as plot size and soil characteristics. While credit access is found not to affect adoption, material support by external agents strongly influences farmers' decision, which may indicate a low initial motivation by land users to undertake such investments on their own.

Soil conservation is a public good as long as its benefits extend not only to the farmers but to the society as a whole, and land degradation needs to be addressed as a societal issue. Therefore, improving the economic attractiveness of recommended SCT needs to be addressed by interdisciplinary research to combine land use options with SCT that are competitive with the prevailing cropping activities. Our results reveal that farmers face important knowledge, institutional, and economic constraints to the adoption of soil conservation practices. This may encourage decision-makers and development organizations in fragile areas not only to promote these new technologies, but also to actively support their adoption by farmers in order to address societal issues of water safety, food security, and sustainable rural development.

Acknowledgement – We thank Holger Seebens, Daniel Müller and anonymous reviewers for valuable comments. Furthermore, we are thankful to Le Thi Ai Van and Dinh Thi Tuyet Van for their contribution in the data collection and to Thai Thi Minh and Andreas Neef for sharing case study findings. We acknowledge the DFG (Deutsche Forschungsgemeinschaft) for funding this research through the special research program SFB 564.

3.8 Appendix: Endogeneity of the land title variable: a test based on Instrumental Variable method[32]

In the paper presented in this Chapter, we studied the impact of land titling on adoption incentives for agroforestry, with household- and plot-level econometric models of adoption (cf. Section 3.5.1 and Section 3.5.2). We found in both models a positive effect of the land title variable on the probability of agroforestry adoption. Both models treat the variable land title as exogeneous. This choice was motivated by endogeneity tests, not shown in the paper but presented in this Section.

There are indeed several reasons to suspect endogeneity of land tenure security in investment models (Place and Swallow, 2000). (*i*) The first reason is that farmers may undertake investments as a way to secure their property. Farmers may plant fences, trees or undertake other investment on their land as a way to enforce their rights in the eyes of officials or neighbors and limit risks of land grabbing. (*ii*) A second argument hinges on the fact that land registered may differ in characteristics with the non-registered land which may also explain higher investment rates. The presence of unobserved differences in attributes between land of different tenure regime, if correlated with the investment decision will produce biased estimates in the adoption model. Take for instance areas where a lease market has established. Endogeneity bias would occur if landlords were to lease out in priority land on which some type of investment are unsuitable. As a second example, take the case of regions in which arable land was not fully exploited at the time of land registration. Land cleared after the registration process will not be formally registered. Access may not be less secured but land quality will differ, and will lead to divergent investment strategies which are not directly linked to the tenure property. The identification of the land tenure effect is blurred by the presence of unobserved differences, which may bias and overestimate our results.

32 This Section was not part of the published article.

Studies such as Besley (1995) in Ghana or Brasselle et al. (2002) in Burkina Faso find evidence of endogenous property rights in land investment models. In addition, both studies find that once this bias is controlled, the estimated effect of property right on investment incentives vanishes to become insignificant, or weakly significant, a result in line with findings from those of Chapter 3.

Are there reason to suspect endogeneity of the title variable in the household and plot-level models of adoption of agroforestry? As explained in Section 3.6.1, all households present in the area at the time of the second registration wave received a land title. Most agricultural land was allocated to farmers with land titles, the rest being kept under community management as a village fund. The land allocation policy, during this second allocation wave was strongly enforced by provincial officials. Qualitative interviews with farmers, community leaders and other local officials revealed that neither farmers nor communities had much to say in this process. There is therefore little reason to suspect that farmers may have undertaken certain types of investments to increase their probability to receive a title, or that those who received a title differed from those who did not in characteristics that may as well affect adoption. The reverse causality effect, i.e. argument (i) enunciated above therefore is unlikely to hold here. This phenomenon was observed mostly in African contexts, characterized by low population density. However, the second source of bias arising from unobserved differences in land attributes between titled land and other land however cannot be disregarded here. It motivates the following analysis.

We test for endogeneity using the Instrumental Variable (IV) method suggested by (Wooldridge, 2002a, p. 477–478). IV method is commonly used in econometrics as a solution to the endogeneity problem (Wooldridge (2002b, chapter 15); Wooldridge (2002a, chapter 5)). In presence of endogeneity, bias arises due to the correlation of one or several regressors with the error term of the structural equation, and can be assimilated to an omitted variable problem. The IV methods extracts the exogenous component of the endogenous variable so as to obtain unbiased estimates in the structural model:

$$y_1 = \alpha_0 + \alpha_1 x_1 + \alpha_2 x_2 + \varepsilon \tag{3.4}$$

$$x_2 = \beta_0 + \beta_1 x_1 + \beta_2 z + u \tag{3.5}$$

where equation 3.4 is the structural equation, y_1 is the dependent variable and x_2 is an endogenous regressor, x_1 are exogenous explanatory variables, and z is the instrument. A good instrument (i.e. which successfully corrects estimates) is a variable that is correlated with the endogenous variable, but is not correlated with the error term of the structural equation, ε. The correlation between ε and u is noted ρ.

As explained by Wooldridge (2002a, p. 477–478) the estimation of an IV model for a binary response model in which the endogenous variable is also binary is nontrivial. The two-step estimates are not efficient in this case, and a full-maximum likelihood estimation procedure must be followed. The author suggests a simpler test of endogeneity based on the work of Rivers and Vuong (1988). We apply the test by estimating the instrumental equation 3.5 in a first stage, where the dependent variable is the variable *title* and the exogenous regressors are X_{3ij}, X_{1i} and z, the dummy variable indicating whether the household was established in 1998. We derive $\hat{u}$, the estimated residual, and estimate the structural probit model y_{3ij} on X_{3ij}, X_{1i}, *title* and $\hat{u}$. The coefficient on $\hat{u}$ provides a test of endogeneity for the *title* variable.

In section 3.6.4 we estimated a plot-level model of adoption of agroforestry, expressed by equation 3.6:

$$y_{3ij} = 1\left[\alpha_1 X_{3ij} + \alpha_2 title + \beta X_{1i} + \varepsilon_{ij} > 0\right], \quad \forall j \in \left[1, T_i\right], \quad j \in \left[1, N^k\right], \quad N^k < N \quad (3.6)$$

where y_{3ij} indicates the adoption status of agroforestry by farmer i on plot j, T is the number of upland plots operated by household i and the sub-sample of households knowing agroforestry as a SCT. We suspect the variable *title*, a dummy variable to be endogenous in this model.

The instrument selected to test for the presence of endogeneity is a dummy variable indicating whether the household was already established in 1998, the year during which the land registration process was carried out in Yen Chau. Households established before this date have thus a much greater probability to have received a title than households established after this date. The variable however is arguably not related to adoption incentives, especially once we control for farming experience, captured here by the variable 'age of the household head'. We indeed look at the extent of adoption ten years after the titles were issued and there is no reason. One reason to suspect that older households have a greater probability to adopt agroforestry would be.

The estimates of the instrumental equation and structural equation with the endogeneity test are presented in Table 3.8:

In the instrumental equation, the selected instrument is strongly significant, indicating that the instrument is valid. The residuals from the instrumental equation is insignificant in the structural equation of agroforestry adoption. As explained above, this leads us not to reject exogeneity of the land title variable in the adoption model, and confirms in turn the validity of estimates in Chapter 3. The same method and instrument were used to test endogeneity of the land title variable in the household-level model and also leads to the same conclusion.

Table 3.8: Instrumental equation and endogeneity test of the land title variable, plot-level probit estimates

	(1) Land title		(2) Adopted agroforestry	
	Coef.	**z-stat**	**Coef.**	**z-stat**
Age household head	–0.011	(1.51)	–0.014	(1.31)
Education+	0.075	(0.30)	0.778*	(1.78)
Actives	0.086	(1.14)	0.105	(0.81)
Expenditure per capita (log)	0.464**	(2.06)	0.238	(0.88)
Poor soil+	0.845***	(4.04)	1.816***	(4.08)
Medium soil+	0.665***	(3.76)	1.550***	(3.46)
Area share	–0.009	(0.02)	1.843***	(4.06)
Steepness+	0.097	(0.70)	0.295	(1.51)
Relative upland size+	–0.165	(0.93)	0.345	(1.30)
Support+	–0.055	(0.31)	1.304***	(4.77)
Credit constraint+	0.106	(0.53)	0.249	(0.94)
HH expects reallocation+	0.299	(0.97)	–0.266	(0.97)
Share villagers expecting reallocation	–0.382	(0.72)	–2.849***	(3.01)
Elevation	0.001*	(1.88)	–0.001	(1.24)
Established 1998+	1.172***	(3.95)	–	
Land title+	–		0.410	(0.44)
$\hat{u}$	–		0.187	(0.53)
Observations		567		567
Pseudo R^2		0.132		0.334
Log likelihood		–245.8		–104.6
χ^2		41.4***		95.6***

z-statistics in parentheses are based on robust standard errors clustered at the household level.
**p* < 0.10, ***p* < 0.05, ****p* < 0.01.

4. Has the reform permitted the emergence of a land market? complementary analyses on the distributional impact of the reform

4.1 Introduction: efficiency and equity effects of land markets

One expected benefit of land registration and titling policies is that it enables the emergence of a land market (Deininger and Feder, 2001). A land market is said to have established when in a given area land is exchanged on a rental or purchase basis, at a market price that equate demand and supply. Both equity and efficiency gains in agricultural production are expected to arise from land markets when those function well (Otsuka, 2007). Land transactions are expected to lead to efficiency gains in the agricultural sector by enabling the transfer of land from less productive to more productive farmers. It also strengthens the labor market by freeing labor force. The emergence of a land market favors the establishment of an off-farm sector, by allowing farmers to sell land and join profitable activities in this sector. Finally, a functioning land market may strongly contribute to reduce poverty. Land and labor are indeed the two main assets owned by rural poor household, enabling them to value these resources on a market will have a positive impact on their wealth, raising the value of their asset, increasing their credit worthiness and risk bearing capacity (de Soto, 2000; Deininger, 2003).

The allocative efficiency and equity outcomes of the land sale and lease markets is nevertheless strongly discussed in the literature, and empirical evidence are very mixed across countries (Otsuka, 2007). The extent and size of such positive effects are in fact strongly depending on the functioning of other markets, and the credit market in particular as well as on the initial land distribution. In presence of credit market failure, the liberalization of a land market may as well lead to increased inequality due to distress land sales from the poor to the rich for instance (Ray, 1998, chapter 12). According to Otsuka (2007) finally land markets are unlikely to rectify inequity of land allocations when distribution is initially unequal.

4.2 Land titling and the land market in Vietnam

The 1993 land law, as explained in section 3.2, granted land users with the right to exchange, transfer, lease, or mortgage or inherit the land, guaranted by the issuance of land titles. One primary motive of the land law was to establish a land market, as a way to induce a more efficient resource allocation, reduce poverty and boost agricultural production (Que, 2005; Ravallion and van de Walle, 2008b, p. 20–34). Several institutional arrangements impeded the establishment of a free land market however: the limitation of lease contract to a three year period, the obligation to register all transactions to local officials, and the resulting administrative red tape. These restrictions were partly motivated by the fear of a rise of rural landlessness (Ravallion and van de Walle, 2008b, chapter 2, chapter 6). In the Northern Uplands in particular, the first land allocation gave rise to tensions between ethnic minority communities and officials, and to local interpretation of the law which also prevented the establishment of a land market (Sikor, 2004). The 2003 land law, in response, sought to "*[encourage] civil transactions on land use rights [...] limiting unnecessary administrative interventions during the implementation of exchanging, transferring, leasing, subleasing, donating land use rights and real property*" (Socialist Republic of Vietnam, 2004, p. 6). It also "*recognizes that land needs to be evaluated by market economic forces*" (Socialist Republic of Vietnam, 2004, p. 6).

National-level studies show that a land market slowly emerged after the 1993 land law. Deininger and Jin (2008) shows that the land sales and rentals that took place after the implementation of the law led to both efficiency and equity gains in the agricultural sector, and that land was transferred from less to more productive farmers. Ravallion and van de Walle (2008b) find evidence that the land market lead to a more equitable distribution, i.e. that it benefited the land poor, those who received little land from the first allocation.

4.3 Land transactions in Yen Chau

In Yen Chau however, the situation contrasts with these national level findings. In 2007, farmers and local officials still complained about administrative red tape in land transactions. These exchanges still had to be registered and approved by local officials (at the village, commune and district level). More strikingly, most households and village heads interviewed in 2007 still believed that land sales were forbidden. Finally, our data show that the land sale market remains thin. We recorded only 16 cases of land sales by households. Interestingly however, the number of plots acquired through purchases recorded in the database is substantially higher (70 cases out of 2410 plots). This imbalance between land sales and purchase is most likely due to an under reporting of land sales in the data. The small number of purchased plots confirms that this practice remains anecdotal in the district. Table 4.1 reports summary statistics on the mode of acquisition of plots recorded in the database. The first part of the table displays the different modes of acquisition of titled and untitled plots. A further distinction is made between paddy fields, fishponds and upland plots as both customary rights and allocation mode differ between the three plot types. Paddy fields are more ancient than upland fields, these last have mostly been cleared during the collectivization (1960–1988) while terraced paddy fields were already under cultivation during the communist period. Thai societies for instance used

Table 4.1: Acquisition modes and contracts of Paddy, Fishpond and Upland plots in Yen Chau

	Paddy	**Fishpond**	**Upland**
Operated with title (Number of plots)	558	214	1,158
Allocated without land title[a] (%)	44.6	5.6	14.3
Allocated with land title (%)	9.9	1.9	8.4
Inherited (%)	19.4	33.2	23.6
Legal purchase (%)	2.2	5.6	1.9
Converted wild land into agricultural land (%)	17.7	43.5	45.4
Given by parents (%)	5.9	9.8	6.4
Exchanged (%)	0.4	0.5	0
Operated without title (Number of plots)	121	26	333
Leased in private (%)	2.5	0	21.9
Borrowed private (%)	39.7	50.0	40.2
Given (%)	9.1	0	5.7
Leased/borrowed from village fund (%)	38.0	11.5	20.1
Unregistered purchase (%)	0	26.9	3.9
Started cultivation without permission (%)	10.7	11.5	8.1

[a] Land has been allocated by the government to the household prior to the registration process, at the time of decollectivization in 1988, the land title was delivered afterwards.

to allocate paddy fields in a rotative system, each field being reallocated after a period of about five years, according to the hierarchical societal structure (Sikor, 2004; Mellac, 2006). Second, while upland fields are reserved for cash cropping activities (maize), paddy fields and fishponds are used mostly to produce rice and grow fish which are mostly consumed by the farm household (while 86.3% of our sample households produce rice, only 11.6% of those derive cash income from this activity).

The first part of the table shows that most of paddy fields were allocated by the state at the time of decollectivization in 1988, and that a large share of these plots (44.6%) were kept by tenants after land registration (in 1993 and 1998). Only a small share of paddy fields was in fact reallocated by the state during land registration (9.9%), and this is also true for upland fields (8.4%). The registration process of land did not result in a complete redistribution of land. Upland fields and fishponds were mostly acquired by farmer by converting wild land into agricultural plots. According to interviewees, the decollectivization in 1988 was followed by a land rush in the uplands, leading to massive deforestation. This trend is well observed here, with 45.5% of upland plots that were acquired through conversion. The same holds true for fishponds. This share is equal to 18% for paddy fields which confirms that those plots are more ancient than the other two types.

As said above, land purchases are extremely rare in Yen Chau. They concern mostly Fishponds. This is supposedly due to stronger tenure rights on these lands and to lower reallocation probabilities. Fishponds are often attached to residential plots, or located very closely to households' homestead. They also need important investments for digging the pond, conveying water to the pond , and maintaining it. These plots are mostly ignored in government's land use plans (data on fish production are rarely collected by commune and district offices since harvest is used essentially for home consumption) and the risk of reallocation is presumably lower than for paddy and upland fields.

The lower part of table 4.1 reports descriptive statistics on contracts under which untitled plots were operated at the time of interview. We classify contract arrangements as follows: a land is considered to be leased from a private household if the use right belonged to a different household than the operating household, if the tenant was paying a rent and/or if a duration of lease was stipulated. In case the land is used without the duration of contract being fixed, or rent being paid, the land is recorded as "borrowed" from a private person. The distinction between "borrowed" and "given" categories is rather thin, but for the sake of clarity here, we differentiate between these two contracts, although in the following graph the two categories are merged into "given". A fourth category separates out the plots that are part of the village land fund (land kept under village control in the first allocation process, to be distributed among newly established households) which are lent or leased to the household. This land can be rented out or simply lent to the households. Practices differ across villages, and across households: those having no land title are usually not required to pay rents, while farmers renting this land to increase their production capacity are often required to pay). The last category finally indicates land that is cultivated "illegally"[33], i.e. operated without land title and without agreement with the landlord, or land that is cultivated but not registered as agricultural land (i.e. forest land or residential land under cultivation). We see that a substantial share of land is still cultivated in such way.

4.4 Distributional effect of the land market

Figure 4.1 displays the reallocation of land that resulted from land transaction in Yen Chau district. This method was inspired by a paper from Boucher et al. (2005), in which the author analyze the analyzing distributional impact of land reforms in Honduras and Nicaragua. The x-axis in figure 4.1 indicates initial land distribution, i.e. the distribution of land under title in 1999. Frequencies are reported on the left y-axis while the right yaxis indicates the area land operated by the household at the time of survey in 2007. All figures are displayed in hectare per capita. The gray histogram displays initial frequency distribution of initial land allocation, i.e. distribution under land title. The thick black line is the 45° line, and represents the hypothetical farm size (operated land) if no transactions had taken place in Yen Chau. The thick red line displays the observed farm size. The gap between the black and the red line indicates that land transaction has taken place, i.e. that land was either given away (leased out or lent away) or that land was accumulated (i.e. leased in or received). In areas where red line is above the black one, farmers received more land than they gave away (net land "receiver"), and in regions where the red line is located below, farmers gave on average more land away than they received (net land "giver"). The intermediary blue lines show under which arrangement was the land exchanged, whether leased (i.e. with fixed rent or duration), given away or received (i.e. no duration and no rent agreed upon). As in Boucher et al. (2005), the lines display non-parametric locally-weighted regression estimates (lowess) with a bandwidth set

33 The appellation "illegal" is ours, and derived from our interpretation of the data. The term was not used as such during interviews.

Fig. 4.1: Land allocation and land transactions in Yen Chau

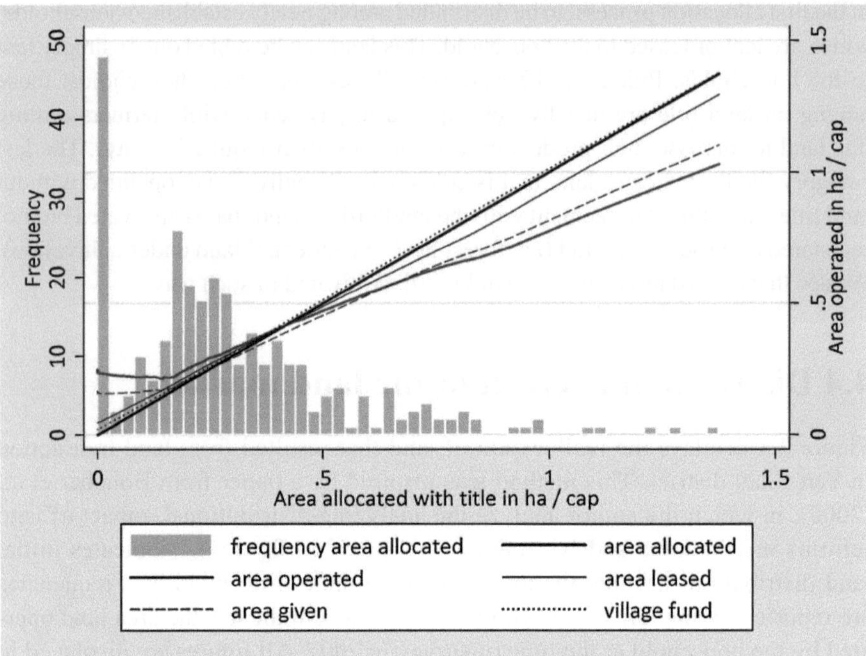

Locally weighted regression estimates, figures are in hectar per capita

at 0.8 enabling a descriptive and visual assessment (see Cameron and Trivedi, 2005, p. 307–311, for a methodological description).

Figure 4.1 displays several striking elements. The first relates to the initial land distribution in the area. The average size of land allocated is equal to 0.297 hectare per capita. The estimated Gini coefficient of initial land distribution in Yen Chau is equal to 0.445, indicating a rather unequal distribution. Considering only household established before 1999, thus eligible to the land titling program, the estimated Gini coefficient is 0.380, and 0.344 if we consider only households holding a title. A substantial share of households, 16.9%, did not receive any land under title. Most of these households were established after 1999, the second wave of land registration in Yen Chau. Referring now to the area of land actually operated by the household at the time of interview figures are somewhat different. The average farm size is equal to 0.345 hectare per capita, and the Gini coefficient is equal to 0.339, slightly lower therefore than the estimated coefficient of initial distribution.

Therefore, since 1999 and 2007, land transactions have led to a more equal distribution of land among households. This can be seen clearly from on Figure 4.1, households initially less endowed in land resource (left side) are net land "receiver" indicating that they have received more land than initially received. As we move towards the right, and as the initial land endowment increases, we see that the relation is inverted, and well endowed households are net land "giver" (they gave land away more than they received). Land exchange and transactions have thus permitted to rectify the inequity of land distribution towards a more equal distribution. Note also that in 2007, our data base contained only six landless households.

The intermediary blue lines indicate under which arrangement were these land exchange made. Figure 4.1 and the thin dashed blue line indicate that most of these exchanges occurred through 'free' contracts, corresponding to land that was lent or given away without a fixed rent or a duration being specified between the two parties. These transfers correspond for a large majority to intra-family transfers. The thin plain blue line indicate that land lease has been taking place, but that the lease market remains very thin in comparison. Finally, readjustment of land distribution through the village fund is anecdotal. We see in addition that these village lands are not specifically allocated to initially land poor households, but to all categories of households. The overall impact on land distribution is anecdotal.

The distributional impact of the reform is further investigated in Figure 4.2, that display the average land holdings of households established before and after the reform implementation in Yen Chau (1999) and details tenure contracts.

As Figure 4.2 shows, household established after 1999 have smaller farm size that those established earlier, their average farm size was 0.31 ha/capita against 0.35 for the other group. This difference however is not significant (Wilcoxon-ranksum test yields a p-value of 0.266). Looking however at the different contract terms, we observe much differences, the younger group holds 0.14 ha/capita of land under land title, against 0.33 for the other group (difference significant at 0.1% of error probability, Wilcoxon-ranksum test). Almost half of land holdings in the younger group comes from free transfers (denoted as given, 0.11 ha/capita on average), and a small share of this land is leased in (0.03 ha/capita, the difference with the first group is however not significant). Land from village funds account for a minor share of households' land holding 0.014 ha/capita in the younger group, against 0.013 for the elder group. The difference between both groups is not significant, showing that village fund land is not allocated to preferentially the households established after 1999. Hence, Figure 4.2 shows that younger households are able to access land mostly through intra-familial transfers, but hold this land under weaker tenure rights than older households.

Fig. 4.2: Average land holding by tenure contract and year of establishment in hectare per capita

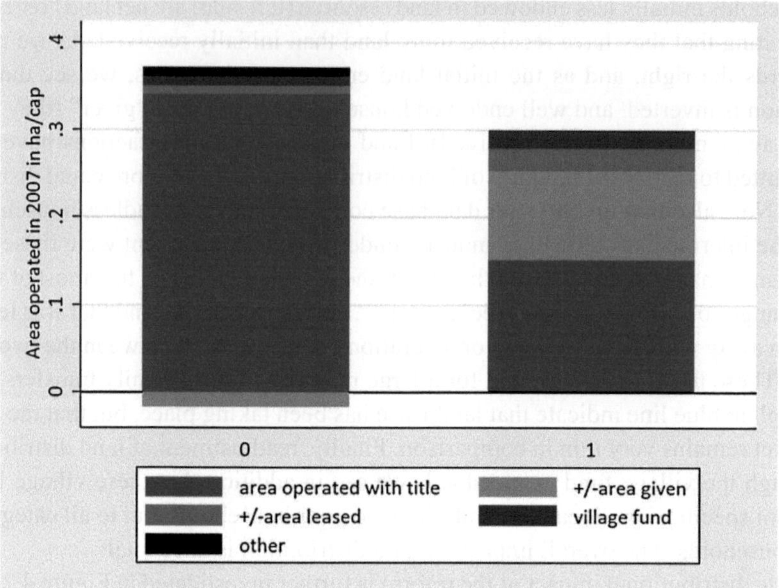

4.5 Conclusion

One objective of the land reforms in Vietnam was to establish a land market as a way to boost agricultural production and reduce poverty. While at the national level a land market has emerged and yielded positive distributional and production effects, our data show a contrasting situation in Yen Chau. There, land sale transactions remain anecdotal, and the lease market is very thin. Reasons to this situation are multiple. First, the little knowledge that many farm households have on the law and granted rights hampers land sale transactions. Second, insecurity raised by reallocations described in section 3.6.1 may as well limit farmers' willingness to engage in sale transactions. The uncertainty regarding the extension of land use rights at the end of farmers' use right term adds to this effect. As we shall see in Chapter 5, the credit market in Yen Chau is functioning relatively well, even though it serves non-poor households better than the poor. Credit constraints may be a limiting factor but is not the main explanation.

We see nevertheless that land transfers have taken place since the land registration. These land transfers resulted in a more equal land distribution than initially induced during the land registration. This adjustment was permitted mostly by intrafamilial land transfer, and a to lesser extent by transactions on the lease market. The redistribution through the village land fund is insignificant. If the current system ensures a relatively good equity of land distribution, the efficiency outcomes and poverty reduction effects of the reform could be greater with the establishment of a land market.

Therefore, one important drawback of the reform is that it deprives the local economy of potentially important efficiency and equity gains. The current system enables almost all farmers in the area to access and cultivate land, and limit inequalities in the short run. In the long run however, negative consequences might arise. The newly established households who have not received land and titles through the reform have only limited possibilities to acquire long term and secure land rights as long as the market does not function well. Young households face higher capital constraints than their elder neighbors and in the current system are doomed to wait for intra-familial transfers or for land out of village fund. As population continues to increase these possibilities will shrink. A majority of villages in our sample had already allocated all land of the village fund, making further expansion impossible. Furthermore, as long as farmers are not be able to sell their use rights on the market, voluntary migrations from the crowded areas to other places offering better employment possibilities will remain limited. If nothing changes, the young generation is doomed to abandon farming and look for other possibilities. Policies are needed to accompany the young generation into this transition, to support the development of an off-farm sector, and help farmers to diversify away from farming.

5. Rural Credit Policy in the Mountains of Northern Vietnam: Sustainability, Outreach and Impact

Abstract

This paper examines the efficiency and equity outcome of Vietnam's rural credit policy in a mountainous district. Using a rich dataset on credit transactions and access from farmers in northern Vietnam, we analyze the credit market, the role played by the two state-owned lenders, and the interaction between the formal and informal sectors. We then assess the impact of the subsidized microcredit program on the welfare of its participants using a propensity score matching approach. Our results reveal a number of inefficiencies that need to be addressed in order to further reduce poverty and support agricultural growth.

5.1 Introduction

Since the decollectivization and the vote of *Doi Moi* program in 1986 leading the transition of the economy from a centrally-planned to a market-oriented system, Vietnam has undertaken a series of reforms aimed at transforming the incentive structure in the rural economy and return farmers to the center of decision making. To this aim, agricultural markets were gradually liberalized, user rights were transferred to smallholder farmers for most of the agricultural land and some of the forest land. As a result, agricultural production and rural economic growth led to a large reduction in poverty throughout the country. These improvements have not been evenly dispersed throughout the country, and mountainous regions, specifically the Central Highland and the Northern Uplands, have lagged behind and experienced lower economic growth and poverty reduction rates in that period. In these mountainous areas, poverty is tightly interlinked with the environmental conditions; the rarefaction and degradation of land resources in particular caused by demographic pressure among other factors remain a major concern and a driver of the poverty trap. Many decisions, in rural areas and ecologically fragile environments in particular, require intertemporal decision making and long horizon planning. Credit access by increasing farmers' risk bearing capacity, and enabling long term investments play a particular role in this perspective (Pender, 1996).

Further growth and diversification of the rural economy, the establishment of more valuable pro-poor agrifood value chains and the promotion of environmentally-friendly economic growth in mountainous regions require the development a competitive rural financial sector.

In rural areas, information asymmetries combined with high levels of uncertainty are the sources of failures in financial markets, leading to regressive credit rationing and thereby creating economic inefficiencies as well as increasing income inequality (Petrick, 2005). These failures and resulting inefficiencies constitute the main rationale for government interventions in the credit sector (Besley, 1994).

Models of intervention have varied over time and across countries. Repressive interventions and subsidization which dominated in developing countries from the 1950s to 1980s have been mostly unsuccessful in addressing market failure and rural poverty and have consequently been abandoned in many countries (Conning and Robinson, 2005). The microfinance 'revolution' in the same period demonstrated that institutional innovations in the sector can successfully enable rural finance institutions to address rural poverty while ensuring their financial sustainability. A new paradigm in rural finance development emerged in the 1980s and advocates for institutional innovations enabling lenders to overcome information asymmetry and enforcement problems as a way to achieve greater outreach and impact, notably among the poor. Under this new paradigm, the role of the state is reviewed, i.e. more focused on promoting of emerging institutions that also serve the poor, rather than on repressive interventions (Lapenu, 2000). Financial sustainability, outreach and impact constitute in this paradigm the three overarching objectives of rural finance institutions (Zeller and Meyer, 2002).

Vietnam, against the dominant trend, has maintained a highly interventionist approach whereby the state continues to control most formal credit supply in rural areas (the totality in our study area). There, government intervention is embodied by two state-owned banks. One is the Vietnamese Bank for Agriculture and Rural Development (VBARD), created in 1990 which acts as a profit-oriented commercial bank supporting the development of rural areas by providing loans to agricultural and non-agricultural enterprises. VBARD's interest rates remain however partly controlled by the state (Duong and Izumida, 2002). The other bank is the Vietnamese Bank for Social Policies (VBSP), which in Vietnam is abbreviated as 'policy bank'. Its mandate is to offer microcredits at preferential interest rates to a targeted population, mainly the poor. The bank transfers its lending activities to party-controlled village-based mass organizations (MO), enabling it to maintain interest rates set below the inflation rate[34]. From an institutional point of view, VBSP's lending system is quite innovative. It makes it a hybrid institution between usual rural micro-banks and village credit funds institutions such as those operating in Thailand (Menkhoff and Rungruxsirivorn, 2011). Yet, the bank does not collect savings and is not financially sustainable; its program remains heavily subsidized by the government. Many rural areas have not yet seen the development of a competitive microfinance sector that is independent from state intervention albeit the socio-economic conditions for the establishment of a vibrant microfinance sector are quite good in comparison.

The objective of the present article is to evaluate Vietnam's rural finance policy performance, by exploring its outreach and impact, and analyzing its role in the existing market. Indeed, one stated rationale of interventions is to offer an attractive formal credit supply as a way to compensate for the inefficiencies of the (private) informal credit market considered unstable, insecure, and unreasonably expensive. This task has proven difficult in many countries including Vietnam. The informal sector remains there a primary credit source for many farmers (Tra Pham and Lensink, 2007; Dufhues and Buchenrieder, 2005; Barslund and Tarp, 2008), an observation that is commonly made in developing countries (Siamwalla et al., 1990; Bell et al., 1997; Kochar, 1997; Mohieldin and Wright, 2000; Guirkinger, 2008; Giné, 2011).

The coexistence of formal and informal lenders in areas where competitive formal contracts are available has raised much interest in the literature in the last decades. Understanding the reasons which underly this phenomenon is essential to identifying sources of policy failure, or preference mechanisms which call for different policy responses. Various explanations have been proposed. First, the persistence of informal lenders is due to an excess demand and credit rationing in the formal sector creating a spillover demand in the informal sector (Bell et al., 1997). This view, however, is challenged by studies such as Kochar (1997) in India showing that conditional on demand, credit rationing in the formal sector is less than usually assumed. In her view, some borrowers face high transaction costs in the

34 The rankings of 1,158 microfinance institutions (MFIs) worldwide in the MIX database (www. mixmarket.org) highlights the peculiarity of the VBSP lending system in the microfinance sector. First, with 7.5 million households in 2009 (about a third of Vietnam's population) VBSP was the largest MFI that year in terms of outreach. Second, the bank operates with low lending costs, US$35.5 per borrower against US $132 on average for other MFIs. These low operating costs are permitted by the reliance on local organizations. Finally, unlike most MFIs, VBSP's real interest rate is negative, so the bank exhibits negative profit margins.

formal sector which exceeds borrowing costs in the informal sector, and rationally turn to the second sector. Boucher and Guirkinger (2007) propose a third explanation based on the risk aversion of poor borrowers. Formal lenders compensate for low information access by requiring borrowers to mortgage collateral, usually their land titles while informal lenders, facing lower information costs are able to substitute collateral through information intensive arrangements. In rural areas, land together with labor constitute the main assets owned by poor farmers, and the risk of losing such crucial assets may convince farmers to turn to more expensive informal lenders[35]. In our view, a fourth explanation lies in the imperfect substitutability between loans of both sectors. Diagne and Zeller (2001) in Malawi, find for instance that demand for credit in a given sector (formal or informal) is only moderately affected by access in the other sector; Mohieldin and Wright (2000) in Egypt find that demand and participation in one sector is independent from participation in the other. Lenders in both sectors differ in their ability to overcome information asymmetries, but also in the nature of contracts offered. Informal loans for instance are more often given in-kind, many transactions are realized within social networks or through existing relationships, and are attached to social or other types of compensations (as in the case of interlinked transactions). Both features may be of greater convenience to the borrower, and not substitutable by formal cash loans. Finally, farmers may seek to diversify their loan portfolio as a way to minimize default risks as well as dependence on any one lender. While even when risk is considered, diversification may not be cost-efficient. Diversification of contracts (with different maturations for instance) can effectively mitigate farmers' default risks in environments with high uncertainty, i.e. when decision-makers are unable to assign subjective probabilities to future risk events[36].

Empirical studies on Vietnam's rural credit policy remain relatively scarce. Duong and Izumida (2002) study the determinants of credit rationing by VBARD, and find positive impact of credit access on household production decisions. Cuong (2008) uses national level data to evaluate the outreach and impact of the government's microcredit program. His results, like those of Dufhues and Buchenrieder (2005), point to the low outreach of the program among the poor. Cuong (2008) finds the program has a positive impact on household income levels and poverty rates at the national level.

35 In an empirical application using panel data from Peru, Guirkinger (2008) finds that transaction costs and risks are factors which prevent farmers from applying for formal loans.

36 Farmers may for instance, borrow from one sector to payback a debt in the other sector.

This article contributes to the literature in two ways. First, it adds to the existing literature on Vietnam's credit policy by describing and analyzing its implementation and impact in a marginal mountainous area. Unlike many existing studies, this paper analyzes Vietnam's policy by studying the credit market in its entirety and not only the formal state-owned lenders. Using a rich primary data set on credit access and transactions described in section 5.2 and combining them with case-study findings, this paper offers a comprehensive overview of the credit market and the performance of different lenders. Such detailed analysis does also carry out more general lessons. Despite several decades of extensive research on rural financial markets in developing countries, many questions remain unanswered, notably on the mechanisms of financial institutions that enable them to reach their objectives, on the functioning of the informal sector and its interaction with the formal sector, and detailed microeconomic evidence is needed to identify policy options likely to resolve market failure and thereby further reduce poverty (Karlan and Morduch, 2010).

Our empirical approach is threefold. First, in section 5.3, we investigate the credit market and its actors. We examine contract terms, participation of borrowers by wealth category, and estimate the determinants of informal interest rates. In section 5.4 we focus on the formal sector and investigate the outreach of both banks. We also investigate in this section interaction between the formal and the non-formal sectors, by estimating the determinants of participation in one sector conditional on participation in the other sector. Finally, in section 5.5, we estimate the impact of the VBSP credit program on participants' welfare using the propensity score matching approach. We discuss the results in section 5.6 and derive important policy recommendations potentially relevant for increasing the outreach, impact and sustainability of the formal sector. We conclude in section 5.7.

5.2 The data

This study takes place in Yen Chau, a mountainous district located in the Northern Uplands region. Despite tremendous economic success in the country over the last thirty years, mountainous areas have lagged behind in the process, and poverty rates among ethnic minorities are significantly higher than in the rest of the country (van de Walle and Gunewardena, 2001; Baulch et al., 2007). Three ethnic groups live in the study area. The Thai were the first settlers and mainly occupy the lowland. They represent 75% of our sample households. The H'mong (14.7%) are late-comers and occupy the uplands. The Kinh (9.3%), the country's main ethnic group (82% of the country's population) settled after the independence from France in 1954. Other ethnic groups account for less than 1% of our sample.

The district is relatively well-off in the province. According to our calculations based on a household expenditure survey using the Living Standard Measurement Method (Grosh and Glewwe, 2000), the poverty rate in 2007 was 16.7% (using the national poverty line and our estimate of per capita expenditure). The majority of the population lives from land cropping. Off-farm opportunities are limited, and livestock production is not well developed. The two main crops are hybrid maize, a cash crop cultivated on rainfed upland, and paddy rice, a food crop cultivated on usually irrigated lowland. Since the paving of the national road crossing the district in 2003, the population benefits from good connections to markets, and maize cultivation remains by far the most profitable activity in the area (Keil et al., 2008).

Land redistribution and titling took place in the 1990s (Saint-Macary et al., 2010). Farmers do not own land but have long term use rights for twenty to fifty years. These land use rights are accepted as collateral by VBARD. Most households established after 1999 were not granted titles from the government but received land from the village or through intra-familial transfer. Consequently, a substantial share of households, 15%, operates on land without land title. A land market has not really emerged, sales transactions are often considered illegal (although officially authorized by law), and the lease market exists but remains very thin. Off-farm opportunities are limited, and the livestock sector was only a complementary income source at the time of study. Land holdings in the area average 0.3 ha/capita, indicating a high land scarcity in this agrarian economy.

Data were collected between 2007 and 2008. A representative sample of 300 households was selected following a two-stage cluster sampling procedure. A village-level sampling frame was constructed encompassing all villages of the district[37], including information on the number of resident households. First, 20 villages were randomly selected using the Probability Proportionate to Size (PPS) method. Next, 15 households were randomly selected in each of these villages using updated village-level household lists. Since the PPS method accounts for differences in the number of resident households between villages in the first stage, this sampling procedure results in a self-weighing sample (Carletto and Morris, 1999).

37 The urban center and four communes located along the border with Laos for which research permits were difficult to obtain at the time of interview were excluded from the sampling frame.

The survey covered a wide range of issues, including land holding, food and nonfood expenditure, income sources and diversification, among others, and a detailed module on households' credit history and access. Data on credit history include information on households' application and complete rejections for large loans in the five years preceding the survey, and the different credit obtained by household members over different recall periods. The use of different recall periods ranging from two months for very small loans to five years for the larger ones enabled us to obtain detailed and accurate information of households borrowing practices. We could record even the very small loans contracted by households to finance daily food consumption in 2007. Farmers were interviewed between May and June, at the end of the lean season, before the first rice harvest and again in December, a few weeks after the maize harvest. The database thus contains detailed information on small and large loans contracted by farmers in 2007, and on larger loans contracted by farmers in the five years preceding the survey. Data was additionally collected on households' credit access, independently from their participation in the market. We use the credit limit approach developed by Diagne et al. (2000). Respondents were asked to estimate the maximum amount the household would be able to borrow from a particular lender, given their own characteristics, current debt level from different lenders. Credit limits were collected separately for each lender type.

5.3 The credit market in Yen Chau

Rural credit markets are known for not functioning like other competitive markets. Transaction costs and information asymmetries in particular are important factors hindering the functioning of this market, resulting in an inefficient allocation in the sense of Pareto (Hoff and Stiglitz, 1990; Conning and Robinson, 2005). We focus on a mountainous region where transaction costs are important and where the fragility of the ecosystem renders climatic hazards particularly severe and frequent (soil erosion notably accentuates the incidence of landslides and floods, Ahlheim et al. (2009)). On the other hand, good intra- and inter-community relationships, whether due to communist or cultural legacies, and the resulting high level of trust in the society create favorable conditions for trust-based transactions such as credit. The literature has indeed established a strong link between social capital (i.e. level of trust, social networks), social cohesion and the performance of rural credit institutions (Zeller, 1998; Cassar and Wydick, 2010; Dufhues et al., 2011).

A conspicuous trend in the data is the high heterogeneity of actors in this market, and the large number of transactions contracted by farmers from all wealth categories. We divide lenders into three sectors and five lender types. The formal sector is composed of the two state-owned banks, the VBARD and the VBSP. The Semi-Formal sector is composed of private or state-owned companies or agencies (Mass Organization, agricultural extension office). The last sector, the informal sector, is comprised of all other lenders. We make a distinction between private lenders (shopkeepers, moneylenders, or any lender without a family relationship or friendship with the borrower) and households' friends and relatives.

5.3.1 Contract terms

Table 5.1 describes the five lender types and the main characteristics of contracts offered in 2007 (Panel A). In Panel B, we report the participation of farmers by wealth tercile, and in Panel C, the market share of each sector measured by the share in households' total debt is shown. In Panels A and B we distinguish three loan types. 'Small' loans are distinguished from the large ones, and the in-kind input loans are separated out. This distinction enables us to directly compare contracts from the formal and informal sectors, the former offering loans of amounts starting at 2 million VND on ($\simeq$ 125 US$). We are also able to observe the participation of households in each market segment. In Panel B, households are separated into wealth groups defined over a wealth index, obtained from the principal component analysis of variables denoting household's wealth status in 2006. We prefer such an index over a measure of per capita expenditure as it is less volatile and less likely to be affected by households' participation in the credit market in 2007. Descriptive statistics of the variables included in the wealth index and their component loadings are described in the appendix of this chapter, Table 5.9.

The number of loans contracted by farmers in 2007 was very large; only 7% of households did not take out a loan. In particular, we recorded a very large number of small loans, six on average per borrower. These loans are mostly in-kind consumption loans from shopkeepers, or small cash loans from friends and relatives. About twothirds of households obtain these small loans. All wealth terciles are concerned, but on average households from the lowest wealth tercile recourse more on these loans. Interest rates reported by respondents are low (6% on average), and most transactions are done within the village where information availability is apparently good.

Table 5.1: Loan characteristics by lender type and participation by wealth tercile (2007)

		VBARD	VBSP	Semi-Formal			Private lenders			Friends & Relatives		
Loan category		> 2M[a]	> 2M[a]	< 2M[a]	> 2M[a]	Inp.[a]	< 2M[a]	> 2M[a]	Inp.[a]	< 2M[a]	> 2M[a]	Inp.[a]
Nb loans in 2007		31	74	14	3	114	1,114	87	216	207	54	21
Amount (Million VND)[b]	Mean	37.6	7.8	0.34	2.7	2.7	0.13	9.0	2.3	0.20	6.6	2.8
	(S.D.)	(48.1)	(3.8)	(0.48)	(0.6)	(2.1)	(0.3)	(32.1)	(2.5)	(0.36)	(13.5)	(2.6)
Interest rate (% p.a.)	Mean	16.1	6.6	13.6	12.9	20.1	6.1	23.8	25.9	0.62	11.0	11.8
	(S.D.)	(1.3)	(2.6)	(13.9)	(7.2)	(15.3)	(16.1)	(14.0)	(19.4)	(3.35)	(12.6)	(9.9)
Duration (nb months)	Mean	22.9	35	6.3	23.7	7.76	4.6	9.6	6.9	4.48	11.2	7.33
	(S.D.)	(10.4)	(7.4)	(3.5)	(12.5)	(2.01)	(1.9)	(5.9)	(3.3)	(2.7)	(8.6)	(1.3)
% Collateral		90.3	1.4	0	33.3	0.9	0.9	5.8	0.9	1.0	0	0
% In Kind		0	0	28.6	0	100.0	92.5	48.3	100.0	39.6	3.7	100.0
% Same village		0	0	0	0	0	64.1	32.2	43.2	72.3	64.8	81.0
Borrowers in 2007 by wealth tercile (%)	Lower	8	22	4	1	24	68	66	66	19	15	1
	Middle	11	23	3	0	38	40	21	48	21	17	8
	Upper	12	26	1	2	32	35	15	37	19	12	10
Pearson's χ^2 test		n.s.	n.s.	n.s.	n.s.	n.s.	***	***	***	n.s.	n.s.	n.s.
% Total current debt	Mean	16.6	22.3		11.6			39.6			10.0	
	(S.D.)	(29.8)	(30.5)		(22.3)			(38.3)			(22.0)	

[a] Loans are classified into three categories: <2M designate small loans below 2 Million VND, >2M designates large loans above 2 Million VND, and Inp. Designates in-kind input loans

[b] In 2007, 1 million VND 62,5 US$

For larger loans of more than 2 million VND, the informal sector is again more popular, with 141 loans, while the formal sector only recorded 105 loans. The largest loans are borrowed from VBARD, other lenders do not differ. Interest rates are significantly lower in the formal sector, as compared with those offered by private lenders. VBSP offers the cheapest loans with a yearly interest rate set at 6.6% well below the inflation rate (12.6% in 2007, according to the General Statistical Office of Vietnam). Loan maturation is longer in the formal sector. In the VBSP case, loan duration is fixed between two and three years, and this period is flexible for VBARD loans, while in the informal sector the maturation does not usually exceed one year. Officially, a formal client is not allowed to contract another formal loan during the borrowing period, but in practice the rule is not always applied. VBARD is the only lender requiring collateral on almost all loans. Other lenders including VBSP, use the information intensive selection to substitute collateral.

Large loans are borrowed by households from all wealth categories, but again, we find a larger share of households from the first wealth tercile among borrowers (64% against 49% of households from the richest wealth tercile). Households from the wealthier tercile group get these loans mostly from the formal sector, while those from the first tercile get such loans mostly from private lenders. The amounts of loans also differ by wealth categories, the richer accessing larger loans, and the poorer accessing more but smaller loans.

5.3.2 Wealth and participation in the credit market

Table 5.2 details participation in the market by wealth terciles. Households from the poorest tercile borrow significantly less than those from the wealthiest one, the first and the second groups however do not differ. The source of credit in both terciles is more diversified than in the third. Households from these two groups borrowed from 2.5 different lenders on average, versus 2.2 in the wealthier tercile. Looking at the leverage ratio, defined here as the ratio of household total debt on their total asset value, we find that the poor borrow more than the other two groups relative to their assets, and thus take greater risks.

We also find differences in credit use by wealth tercile. The poor use a smaller share of credit for financing agricultural activity, and this is also true for agricultural input which constitutes the first borrowing purpose in all groups, they spend however a larger share of credit for financing transportation assets, food and emergency expenses (the absolute value is also significantly larger).

Table 5.2: Participation in the credit market by wealth tercile (2007)

	Lowest		Middle		Highest	
	Mean	**(S.D.)**	**Mean**	**(S.D.)**	**Mean**	**(S.D.)**
Participation						
Share of borrowers	0.96	(0.20)	0.93	(0.26)	0.90	(0.30)
Amount borrowed in 2007[d] (total)	9.03	(7.18)	11.30	(12.20)	22.30	(54.80)
- from VBARD	0.87	(2.93)	2.56	(8.82)	9.4*	(33.00)
- from VBSP	1.54	(2.98)	2.45	(5.09)	2.20	(3.90)
- from Semi-Formal lender	0.73	(1.41)	1.37	(2.21)	1.3*	(2.18)
- from Private lenders	4.82	(5.28)	3.48	(4.09)	6.95	(32.10)
- from Friends and Relatives	1.08	(2.88)	1.41	(3.24)	2.46	(10.90)
Number of different credit source	2.56	(1.10)	2.48	(1.25)	2.23*	(1.11)
Value of assets[a] (Million VND)	25.6	(19.00)	38.5*	(22.70)	50.7*	(43.90)
Leverage ratio (debt/asset)	0.57	(0.83)	0.32*	(0.32)	0.35*	(0.53)
Credit use, % amount used on						
Agricultural production	48.7	(33.40)	66.0*	(32.50)	62.2*	(38.90)
Agricultural input	41.9	(32.10)	57.7*	(34.20)	55.0*	(39.70)
Non-Farm business	4.5	(17.03)	3.7	(16.75)	7.7	(24.21)
Transportation	8.8	(22.29)	3.3*	(14.03)	2.7*	(15.21)
Durables or housing	4.4	(13.60)	2.6	(10.60)	1.41	(7.59)
Food, social event or other emergency	25.1	(26.50)	11.2*	(20.60)	11.2*	(22.90)
Credit source for agricultural input loans (%)						
Formal	9.85	(26.50)	5.56	(19.40)	8.53	(24.50)
Semi-Formal	20.60	(37.90)	37.7*	(46.10)	33.4*	(44.60)
Informal	69.50	(43.90)	56.70	(46.30)	58.00	(46.60)
Interest rate paid (% p.a.)[c]						
All loans	20.8	(12.92)	16.7*	(10.72)	15.5*	(14.55)
Non Formal loans	23.5	(12.49)	17.7*	(11.63)	18.0*	(16.38)
Agricultural input loans	24.6	(18.04)	19.6*	(12.27)	19.4*	(17.39)
Consumption loans[b]	43.5	(53.00)	25.3	(78.20)	8.0*	(21.46)
Credit access						
Credit limit VBARD[d]	4.6	(6.37)	11.0*	(12.69)	22.9*	(33.17)
% borrowed	27.7	(45.20)	38.4	(49.00)	38.2	(48.90)
Credit limit VBSP	4.4	(4.01)	7.3*	(6.41)	7.6*	(6.28)
% borrowed	45.3	(49.40)	54.0	(48.50)	55.2	(49.10)
Credit limit Semi-Formal	1.5	(2.25)	4.7	(20.39)	3.2*	(4.18)
% borrowed	41.7	(44.00)	38.2	(40.70)	34.2	(40.70)
Credit limit Private lenders	8.6	(6.93)	13.0*	(11.35)	19.6*	(38.66)
% borrowed	52.2	(33.10)	29.0*	(30.30)	21.6*	(28.30)
Credit limit Friends and relatives	3.2	(4.21)	10.4*	(15.52)	11.9*	(19.87)
% borrowed	26.6	(32.70)	11.0*	(20.50)	10.9*	(23.90)

[a] Includes house value, durables possessed and value of livestock

[b] Includes loans for the purchase durables, housing, food and other emergency expenditure

[c] Weighted mean of interest rates paid on loan. Weight accounts for loan size within each category.

[d] In Million VND. In 2007, 1 Million VND 62.5US$

*Indicates a significant difference to 1st tercile at 5% level of error probability

Looking now at how households within each category finance their agricultural input, we find that while the purchase of agricultural input is the first borrowing purpose, only a few households finance these purchases with formal loans. The semiformal and informal sectors are the first financial source, and the poor use the latter more than the semi-formal on average compared with the two other groups. Finally, looking at contract terms obtained by households in each category, i.e. the interest rate paid for an average loan, we find that the poor pay higher interest rates than others. This holds true when taking into account only non-formal loans, agricultural input loans, or consumption loans.

In the last part of the Table we display credit limits reported by respondents from each lender types. The concept of credit limit was introduced by Diagne et al. (2000) and tested empirically by Diagne and Zeller (2001). It approximates the maximum amount a household is able to borrow from a particular lender according to its own characteristics. Households decide how much to borrow depending on their need and their credit limit in different sectors Diagne and Zeller (2001). Note that the credit constrained households are not necessarily those borrowing the totality of this limit. A farmer is constrained if his credit limit is not sufficient to cover his need. Credit constrained households, may thus have a positive unused credit limit. The decision on how much to borrow is also part of farmers' risk coping strategy. Indeed, the amount that is not borrowed today will be available in the future in case a shock occurs. We report the share credit limit that was borrowed at the time of interviews.

We find that for all lender types poorer households have lower borrowing capacities, including VBSP. These amounts are, except for VBSP, between two and three times smaller in the first group than in the last one. Looking now at the share of credit limit borrowed, we find no differences among households in the formal and semi-formal sector. We find however that the poor use a larger share of their credit limit from both private lenders and from friends and relatives. Additionally, data shows that many farmers do not borrow from the formal sector while they have positive access, despite lower interest rates in this sector. In the case of VBSP, as we shall see in the next section, it is unclear to which extent households can decide to borrow (loans are disbursed in a top-down approach, and the ability of farmers to borrow depends on the amount allocated by the bank to their village, and their ability to get votes from their neighbors, see section 5.4.1). In the case of VBARD, such restrictions do not exist, and farmers can freely decide whether to borrow or not, depending conditional on their access.

The market appears to be rather fluid: most households are able to borrow in order to finance consumption or agricultural inputs. We observe a high heterogeneity of lenders and contracts. Informal lenders, compared with both banks are able to provide a variety of contracts to farmers from all wealth categories. The ability of informal lenders to use local information enables them to adapt contracts to household demand better than formal lenders can. Comparatively, VBSP offers very standardized loans. VBARD loans are more flexible, but partly restrained by political restriction (capped interest rate).

Institutional heterogeneity is considered a likable feature of credit markets, raising competition among lenders and increasing credit access (Zeller, 2003). The market however suffers imperfections that seem most detrimental to the poorest. Like in many developing countries[38] rich households borrow more and pay lower interest rates than the poor. This is also true considering only the non-formal market. The poor on the other hand are forced to take higher risks to cover their need for agricultural input and consumption, and face higher interest rates.

We explore this issue further by running a loan-level model estimating the determinants of interest rates in the semi-formal and informal sectors. We use a Tobit model to account for the many zeros contained in the database. We find that interest rates increase with the loan size and decrease above a certain level (6.7 million VND). As expected, loans for which collateral is provided have lower interest rates. In-kind loans are cheaper than cash loans presumably because of lower transaction costs, and the weak probability of moral hazard in in-kind transactions, unless when those concern agricultural input transactions. For the latter, transportation costs of input into villages might explain these higher costs.

The interest rate is set higher when it is the first transaction between a particular lender and a borrower, indicating as predicted by theory (Stiglitz and Weiss, 1981; Hoff and Stiglitz, 1990) that information and trust play an important role.

---

38 Banerjee (2003) reviews the literature and enumerates six common salient features: 1. important gap between the deposit and lending rate; 2. high variability of interest rates for similar loan transactions; 3. low default rates; 4. production and trade are the first borrowing purposes even when interest rates are very high; 5. richer people borrow more and pay lower interest rates; 6. bigger loans are associated with lower interest rates. Most of these features apply to the credit market of Yen Chau. Formal deposit interest rate in VBARD varies between 0.25 to 0.74% monthly (3–9% yearly), thus a gap of 7–13% with the bank's lending rates. We do not find a negative relationship between loan size and interest rate in the formal sector, but a quadratic relationship as shown in Table 5.3. Finally, the default rates, only three households lost their collateral through loan transaction in the past five years. Overdue debt is more important, we find a rate of overdue loans (from six months overdue) of 6.6% in and 7.7% for VBSP loans. Rates are lower for the informal sector, but also more difficult to estimate.

Everything else being unchanged, we find that interest rates decrease the higher a household's wealth level and labor capacity, and the larger the area of paddy per capita held by the household. Indicators of wealth and productive capacity are used by a lender as indicators of households' creditworthiness. Lenders may compensate for the greater default risks in transactions with poorer farmers by charging higher interest rates. Wealthier households may also have greater information access and richer social networks that enable them to borrow at cheaper rate. Finally as expected, we find that *ceteris paribus* friends and relatives charge lower interest rates, but that shopkeepers charge more. Higher level of trust and the possibility to exert social pressure on borrowers enable family members to lend at lower rate.

5.4 Formal lending in Yen Chau

In the previous section, we have seen that the formal sector remains a secondary source of credit, despite offering advantageous contracts. A large share of farmers having access to such loans continues to rely on the informal sector where they borrow at higher interest rate. As mentioned in the introduction, this phenomenon has been commonly observed, but the explanations diverge. We analyze in this section the lending policies and outreach of both banks. We estimate econometrically the determinants of demand and access to each program and thus the outreach of both banks. Finally, we explore the interaction between the formal and non-formal sectors by jointly estimating determinants of participation in each sector.

5.4.1 Lending procedure

The Vietnam Bank for Agriculture and Rural Development (VBARD) is one of Vietnam's four commercial banks created in 1988 after the first financial reform. Apart from a few subsidized programs targeted at poor households, most of its activity focuses on commercial banking. Nowadays, it is the largest commercial bank operating in the country, but remains a secondary source in Yen Chau, as seen above. The Bank's interest rates are partly determined by the State Bank of Vietnam (Duong and Izumida, 2002). Local banks adjust interest rates within this defined range. Most VBARD loans require the provision of collateral, which in most cases is a household's land title, and in rare cases some other valuable good. The bank is located in the district's town and all operations are done there, loan

Table 5.3: Tobit estimates of loan interest rates in the non formal sector (2007)

	Mean	S.D.	Coef.	z-stat
Interest rate (% p.a.)	10.43	(17.80)		
Loan characteristics				
Amount (log)	–1.98	(1.98)	10.427***	(3.84)
Amount squared	67.94	(2275.78)	–0.001***	(4.27)
Duration (months)	5.50	(3.39)	0.469	(0.80)
Collateral+	0.01	(0.11)	–21.065***	(2.80)
Guarantor+	0.04	(0.19)	–5.943	(1.35)
In-kind input loans+	0.19	(0.40)	12.671***	(2.92)
In-kind food+	0.57	(0.50)	1.534	(0.25)
Other in-kind+	0.09	(0.29)	–24.685***	(3.47)
In village+	0.58	(0.49)	3.549	(0.71)
First transaction+	0.15	(0.35)	6.357*	(1.80)
Borrower characteristics				
Wealth index	–0.04	(1.00)	–4.554*	(1.86)
Number of actives in HH	2.53	(1.24)	–3.559**	(2.07)
Paddy area 2007	3.56	(3.30)	–1.819***	(2.74)
Upland area 2007	29.61	(19.05)	–0.073	(0.67)
Land title+	0.84	(0.37)	9.44	(1.25)
Lender characteristics				
Semi-formal lender+	0.07	(0.26)	5.82	(0.85)
Friend & Relative+	0.11	(0.32)	–24.519***	(4.87)
Shopkeeper+	0.71	(0.45)	8.057	(1.34)
Moneylender+	0.01	(0.08)	1.175	(0.18)
Trader+	0.06	(0.23)	3.415	(0.34)
Constant			6.59	(0.75)
σ				28.900***
Observations				1583
pseudo-R^2				0.112

+ indicates dummy variables (yes=1, no=0)
t-statistics are based on robust standard errors clustered at the household level
Frequency weights account for the different recall periods by loan types
* p>0.1, ** p>0.05, *** p>0.01

application, delivery, repayment, etc. In rare cases the bank's staff may travel to the very remote villages to carry out transactions. Applicants fill an application form where they detail their income, projects and other information. The form has then to be signed by the village head to confirm borrowers' residence and liability.

The Vietnamese Bank for Social Policy (VBSP)'s lending procedure is relatively complex, and follows a top-down approach summarized below (Kim Anh, 2010). In a first stage, the bank decides, in line with policy objectives, to which commune, under which program and which amount to allocate. The commune's poverty reduction board composed of village heads, and the commune leaders of four mass organizations (MO) (Women Union, Farmer Union, the Youth Union and the Veteran Union), receive the bank's instructions and select villages and the MO in charge of arranging the transactions[39]. At the village level, the selected village MO form a group of borrowers. Borrowers are usually selected during a VMO meeting, during which candidate households expose their projects and needs. Borrowers are selected sometimes through votes or directly by the village MO leader. The list of selected borrowers is then submitted for approval to the VBSP. Once approved, the VBSP staff travels to the village and carries out the transaction. The village MO in most cases is the lending group lender, and the village head acts as guarantor by signing a statement that the selected borrower is liable.

The bank lends under various programs all serving different objectives. The 'poor program' is the main one, and captures 87% of bank's resources, according to the local branch. Other programs aim to foster employment opportunities or to encourage education programs. Farmers eligible for the first program are those that have been identified as poor by the commune. The classification of households into wealth classes is done every year, following the method designed by the Ministry of Labor, Invalids and Social Affairs (MOLISA). Data on household income level are collected every two years. Those with a per capita income level below the rural poverty line (200,000 VND, $\simeq$ 12.5US$ at the time of interview) and with low asset value, are classified as poor and are thus eligible for VBSP's program.

---

39 Districts encompass communes which are composed of villages. The mass organizations are associations created by the state during the communist era to engage civil society in the country's political project. Twenty years after the decollectivization, mass organizations continue to play a major role. The six unions are present at all levels of the society, from the national to the village level, and accompany the implementation of state policies.

As in many MFIs, group-lending is used to substitute collateral requirement by transferring monitoring and enforcement tasks to the group. However in the VBSP case, borrowers from the same village are not liable for the repayment by their neighbors, but a high default rate in the village compromises the ability of other villagers to obtain future loans, giving the village MO a strong incentive to ensure the repayment of each borrower.

5.4.2 Demand and participation in the formal sector

We estimate the determinants of participation in one or the other program conditional on demand, using a bivariate approach with partial observability. All households may not have a demand for a formal loan, and those that have a positive demand may not be similar in their probability of receiving a loan. We apply a two-stage procedure to address this selectivity issue, using a Heckman specification for binary response model (Wooldridge, 2002a, p. 477–78 and p. 570–71). This two-stage approach enables us to disentangle determinants of households' demand for a certain program, from factors influencing a bank's response. A household has a demand for a loan if he applied for a loan, or did not apply for fear of being rejected. As positive demand, we also count cases of households that obtained a formal loan that was not borrowed in their name. The second dependent variable, *granted* is equal to 1 if the household received a loan from the bank during the period considered. The model is written as follows:

$$y_i^d = 1[\beta_1 X_{1i} + \beta_2 X_{2i} + u_i > 0] \qquad \forall i \in [1, N] \qquad (demand) \tag{5.1}$$

$$y_i^g = \begin{cases} 1[\gamma X_{1i} + \upsilon_i > 0] & \text{if} \quad y_i^d = 1 \\ 0 & \text{otherwise} \end{cases} \qquad (granted) \tag{5.2}$$

where y_i^d denotes demand status of household i in a credit program, y_i^g denotes a bank's decision to allocate a loan to a household i. X_{1i} is a vector of household characteristics that influence both demand and a bank's response and X_{2i} are characteristics expected to influence demand without affecting a bank's response. We assume the error terms, u_i and υ_i to be jointly normally distributed. The correlation between u_i and υ_i is noted ρ.

We use data from demand and participation between 2005 and 2007. As explained above, VBSP loans have a duration of three years, and officially no other formal loans can be obtained by the household during that period (in practice however, this is not always the case). In theory at least, each household has the possibility to participate in the VBSP program once every three years, and this is the period considered in the model.

Regressors account for household demographics, education level, asset and land holdings before 2005 in addition to household poverty classification[40], and the possession of land title. A dummy variable indicates whether the household's main occupation is farming. Two variables measure participation of household members in village organizations. Variables are weighted household level mean, in which weight accounts for the degree of participation of each member (with the highest weight set for leaders, and the lowest for the non-actives). We distinguish between 'political' and 'non-political' organization, the first one referring to state-created associations, such as mass organizations and the communist party, and the second refers to all other types of associations (civic associations, cultural and leisure groups, professional organizations, etc.). These variables are likely to affect participation and credit access for two reasons. First, households may gather information on credit programs and the application procedure in both banks. Second, as explained above, the VBSP lending program largely depends on the village MO to select borrowers. Membership, and the degree of participation in a village MO are thus likely to affect the probability of access to a VBSP loan, through information access and bargaining position. The ability of household members to speak Kinh, the language in which VBARD transactions are done is expected to affect both demand and access in the VBARD case.

Village level variables and geographic attributes are expected to affect both demand and access to credit through the availability of other credit sources and through transaction costs. We include a variable measuring the traveling distance to the city, the distance to the closest market. In the VBSP model, we control for classification of communes into economic zones, and for villages' relative wealth within communes[41]. Village and commune classification may enter into account in VBSP's definition of objectives and thereby affect its targeting strategy. Second, the availability of credit in one's village results from negotiations taking place between communes and villages to attract VBSP funds every year. A household

40 Households are classified into five wealth categories every year. The classification method elaborated by MOLISA is based on the biannual calculation of households' income, and the inclusion of other asset-based indicators.

41 The economic classification of communes into Zones I, II and III is a formal classification used by the state to target poor areas with anti-poverty programs. Communes in Zone III are the poorest. This classification is based on several indicators, such as infrastructure, poverty incidence, etc. The relative classification of villages is a self-assessment by village head.

is able to apply for a VBSP loan only when funds are available in its village. The demand for liquidity is expected to depend on a households' income source. We control for the share of farm income and the share of wage income in households' total cash income. We use lagged values denoting households' situation in 2002 to avoid endogeneity issues. Finally a dummy variable indicates whether the household experienced a serious and negative income shock in the five years preceding the survey. The variable accounts for complete crop failures, the loss of a cow or buffalo, the death or departure of a working adult, or the occurrence of a wedding for which the household had to pay dowry. See Table 5.10 in the appendix of this chapter for a description of all household and village level variables.

A check of the correlation matrix between all regressors and of the variance inflation factors indicates no cause for concern regarding multicollinearity. Estimation results are reported in Tables 5.4 (VBARD) and 5.5 (VBSP). In both Tables, columns (1) and (2) display two-stage estimation results, while column (3) presents the marginal effects on the joint probability to demand and receive a formal loan, that is, the estimated effect of an incremental deviation in the explanatory variable from its mean value on the probability for a sample household to participate in one credit program. Marginal effects are reported in percentages.

Demand and access to VBARD credit program

The overall model is significant, and the Wald test on the significance of ρ indicate that both stages are independents. Hence, those who have no demand for VBARD loans do not differ from others in their probability of receiving a loan. The older the household head, the less likely the household is to be granted a VBARD loan conditional on demand. While female-headed households are more likely to demand VBARD loans, male headed households are more likely to get a VBARD loan. Overall, the age of the household head appears to have no impact on household participation. Household demand increases with its paddy and upland farm size, but decreases beyond a certain level (4.5 and 43.8 acre per capita, respectively, corresponding to the 76th and the 87th percentile, i.e. the relation is positive for the majority of the sample). The value of durables, and household poverty status affect neither the demand nor one's access to VBARD loans. Participation in a village organization, which captures information access positively affects households' probability of receiving a loan conditional on demand, but the marginal effect on the probability of becoming a client is insignificant. Households whose members are able to speak Kinh tend to demand more credit; we find a positive impact of the variable on the probability of becoming a client.

Table 5.4: Determinants of demand and participation in VBARD credit program (2005–2007). Probit with selection

	VBARD					
	P(demand>0)		**P(granted=1)**		**Marginal effect P(demand>0, granted=1)**	
	(1)		**(2)**		**(3)**	
	Coef.	**t-stat**	**Coef.**	**t-stat**	**dydx x 100**	**z-stat**
Age head	0.0039	(0.46)	–0.0368*	(1.77)	–0.073	(0.36)
Male head+	–0.5394	(1.45)	1.9031***	(3.41)	–4.991	(0.50)
Education	0.0507	(1.27)	0.0081	(0.09)	1.363	(1.33)
Active females	0.0237	(0.18)	0.0565	(0.20)	0.888	(0.24)
Active males	–0.0173	(0.14)	0.217	(0.70)	0.584	(0.19)
Dependants	0.0861	(1.21)	–0.0121	(0.08)	2.188	(1.18)
Paddy area	0.1542	(1.62)	–0.3238	(1.49)	2.477	(0.95)
Paddy area squared	–0.0180**	(2.11)	0.0206	(1.27)	–0.371*	(1.72)
Upland area	0.0276**	(2.01)	0.0731*	(1.88)	1.070**	(2.41)
Upland area squared	–0.0003**	(2.12)	–0.0010**	(2.07)	–0.013**	(2.39)
Durables (log)	–0.0345	(0.23)	–0.2453	(0.92)	–2.072	(0.50)
Poor 2004–2006+	–0.0345	(0.15)	–0.138	(0.27)	–1.558	(0.26)
Land title+	0.367	(1.12)	2.5823*	(1.86)	21.901**	(2.31)
Participation – P	0.1027	(0.57)	0.3682	(0.87)	4.436	(0.88)
Participation – NFP	0.2091	(0.92)	1.6560**	(2.41)	13.360*	(1.89)
Kinh language	0.0165**	(2.43)	–0.0184	(1.58)	0.342**	(2.18)
Distance to city	0.0004	(0.14)	–0.0036	(0.37)	–0.006	(0.08)
Distance to market	0.0238*	(1.65)	0.001	(0.03)	0.627	(1.63)
Wage income (share)	–0.0267***	(3.41)	–		–0.696***	(2.97)
Farm income (share)	–0.0291***	(4.60)	–		–0.758***	(4.08)
Income shock+	0.5561*	(1.85)	–		14.510*	(1.65)
Observation						288
Censored						188
Log-likelihood						–181.62
χ^2						102.3***
p-value Wald test (ρ=0)						0.617

z-statistics are based on robust standard errors
\+ indicates dummy variables
* p>0.1, ** p>0.05, *** p>0.01

We find that demand for VBARD loans decreases with the degree of specialization into farming, and with the share of wage income in households' total cash income. Consequently, the probability for households specialized into farming and engaged in wage activities to receive a VBARD loan is also lower. This is an indication, as already observed in section 5.3.2 that VBARD loans are directed more towards financing nonagricultural activities than agricultural investment[42].

To summarize, results indicate a selection of borrowers into the VBARD program that results both from a self-selection and a selection by the bank itself. Everything else being equal, we do not find that poorer households are less likely to participate, but that households with a larger farm size have a higher probability of becoming clients. Households which hold a land title are much more likely to take part in VBARD's credit program than others. Finally households that are more engaged in the non-agricultural sector are more likely to participate in the program.

Demand and outreach of VBSP credit program

We turn now to the estimation results of demand and participation into the VBSP credit program (Table 5.5). As explained earlier, the objective of this credit program is to increase credit access to the poor by offering low-risk credit to those identified as poor. We suspect that the complexity of the lending procedure, and notably the involvement of community undermine the bank's capacity to achieve a good poverty outreach.

As in the previous model, the significance level of the model is satisfactory. A Wald-test on ρ leads us to conclude that both stages are independent of each other. We find that the probability of having a positive demand for a VBSP loan increases with education level and the number of active males in the households. It decreases with the presence of active females. The probability to receive a loan conditional on demand increases with the number of dependents in the household, but decreases with the age of the household. It increases with a household's upland

42 Although the two explanatory variables are measured with lagged values, a risk of endogeneity bias persists. Household participation in the off-farm sector and wage employment could be correlated with uncontrolled factors present in the error term, for instance here the extent of past transactions between the bank and the farmer. The likelihood that one receives a VBARD loan today also likely to be positively influenced by the extent of past transactions. This is checked for by including a control dummy variable indicating whether past transactions occurred between the bank and the farmer. This operation leaves the results unchanged, we conclude that the effects estimated here are not affected by past transactions.

Table 5.5: Determinants of demand and participation in VBSP credit program (2005–2007). Probit with selection

	VBSP					
	P(demand>0)		P(granted=1)		Marginal effect P(demand>0, granted=1)	
	(1)		(2)		(3)	
	Coef.	**t-stat**	**Coef.**	**t-stat**	**dydx x 100**	**z-stat**
Age head	–0.0017	(0.25)	–0.0460*	(1.93)	–0.344	(1.51)
Male head+	0.2515	(0.81)	0.9167	(1.35)	13.605	(1.39)
Education	0.0756**	(2.00)	0.1206	(1.40)	3.114**	(2.52)
Active females	–0.2783**	(2.07)	–0.3859	(0.82)	–11.096**	(2.42)
Active males	0.3111**	(2.50)	0.8328	(1.52)	14.933	(3.68)
Dependants	0.0086	(0.15)	0.3874**	(2.01)	2.711	(1.32)
Paddy area	–0.0100	(0.28)	–0.1722	(0.70)	–1.396	(0.84)
Paddy area squared	0.0004	(0.67)	0.0042	(0.25)	0.040	(0.39)
Upland area	0.0007	(0.06)	0.0698***	(2.69)	0.462	(1.15)
Upland area squared	0.0000	(0.23)	–0.0004**	(2.25)	–0.002	(0.36)
Durables (log)	–0.1859	(1.43)	–0.3681	(1.07)	–8.107*	(1.89)
Poor 2004–2006+	–0.2053	(1.01)	–0.8085	(1.44)	–11.488*	(1.85)
Land title+	0.2329	(0.89)	0.0454	(0.10)	7.535	(0.92)
Participation – P	–0.1258	(0.75)	0.6554*	(1.82)	0.216	(0.04)
Participation – NFP	0.0781	(0.37)	0.1221	(0.32)	3.200	(0.46)
Zone I+	–0.2319	(1.18)	–0.2155	(0.52)	–8.575	(1.37)
Zone III+	–0.7597**	(2.02)	–6.5217**	(2.12)	–64.760***	(4.15)
Richer village+	0.2374	(1.27)	0.0845	(0.19)	7.922	(1.30)
Poorer village+	0.4095	(1.41)	5.8789**	(2.52)	49.807***	(3.80)
Distance village center	–0.1900	(0.77)	–0.5080	(1.23)	–9.116	(1.31)
Wage income (share)	–0.0058	(0.83)	–		–0.181	(0.80)
Farm income (share)	–0.0079	(1.28)	–		–0.247	(1.26)
Income shock+	0.0494	(0.20)	–		1.539	(0.20)
Observation						288
Censored						144
Log-likelihood						–216.35
χ^2						883.4***
p-value Wald test (ρ=0)						0.664

z-statistics are based on robust standard errors
+ indicates dummy variables
* p>0.1, ** p>0.05, *** p>0.01

farm size, but decreases beyond the level of 82 acre per capita (99th percentile), the relation is thus positive for most of the population. Households located in a poorer economic zone (Zone III) have a lower demand, and also a lower probability of being granted a loan conditional on demand. This result is surprising, as poor communes are officially targeted in the implementation of anti-poverty programs. Two villages in the sample belong to this poor zone, two particularly remote and poor H'mong villages. In the current VBSP procedure these villages may face difficulties attracting public funds, while richer villages in poor communes, not represented in this sample may benefit from greater access. We find that conditional on demand, villagers from poorer communities have a higher probability of getting a loan, indicating a certain form of targeting at the commune level.

Looking at the last column and at the estimation of marginal effects, we see that better educated households are more likely to become clients. This probability decreases with the number of women in the household, but increases as the number of active male increases. We would expect a positive effect on both variables, as they are both indicators of household labor availability. We find that the more active females in the household the lower the probability of having a positive demand for VBSP. One possible interpretation is that females have access to additional credit sources, through the women's union or through social networks, which reduces their demand for VBSP loans and thus their participation in the program. This indicates, nevertheless, that the microcredit program of VBSP is not gender-sensitive and does not, as many other MFIs do, seek to target women as a way to reduce gender inequality, increase welfare impact and repayment. Our data show nevertheless a much higher share of women among VBSP clients (33.8%) than among VBARD clients (11.6%). If participation decreases with households' wealth level (proxied by the value of durables owned in 2004), those classified as poor during that period, had a significantly lower chance of becoming a client, an indication that the bank may be mistargeting.

We explore the targeting performance of VBSP further, by looking at the probability of a household classified as poor in period $t-1$ applying for and obtaining a VBSP loan in time t, using this time all information available on participation between 2002 and 2007. We expect to find over the time period a bias in loan allocation towards households classified as poor in the previous period. Results are reported in Table 5.6, and indicate otherwise. The poor have a lower probability than non-poor of obtaining a VBSP loan in the year following their classification. We find furthermore that this is not due to a lower probability of application, but to the fact that the poor are more likely to see their application rejected by VBSP. Results also indicate a low depth of outreach, the poor are less likely than

Table 5.6: Poverty outreach of VBSP (2002–2007)

		Classification in $t-1$		χ^2-test for equal distribution
		Poor	Non-Poor	p-value
Obtained a VBSP loan in *t*	n	31	157	
	% within class	9.75	15.05	0.016
Applied for VBSP loan in *t*	n	45	174	
	% within class	14.15	15.43	0.575
Obtained a loan while applied in *t*	n	31	157	
	% within class	68.89	90.23	0.000

non-poor to access VBSP loans, and the share of poor that received such loans is also low (less than 10% of poor households on average per year). In addition, we find that VBSP loans do not reach households without alternative access to formal credit. In 2007, 40.9% of VBSP clients had already borrowed a loan from VBARD between 2002 and 2006, and 71.8% had a positive credit limit from VBARD, meaning that they could have accessed a VBARD loan instead.

The current loan allocation results in a distribution of VBSP loans that is worse for the poor than what would be produced by a random allocation of loans in the population. This poor targeting performance of VBSP results in a major leakage of state resources as well as a strong undercoverage of the targeted population. Based on the data, we calculate a leakage rate[43] of 54.1% and an undercoverage rate[44] of 90.2%.

These results point to the limits of the community-based targeting system used in Vietnam for the implementation of many anti-poverty programs.

Theoretically however, advantages are numerous (Conning and Kevane, 2002). First, it enables the programmer (here the VBSP) to make use of rich information available at the community level to identify at a low cost the population in need and to achieve a better targeting. This enables VBSP to partly overcome information asymmetry and risks of adverse selection (i.e. the selection of 'bad' borrowers with

43 Expressed as the share of non-targeted recipients (non-poor clients) in the total targeted population (the poor).

44 Expressed as the share of targeted non-recipient (poor non-clients) over the total targeted population (the poor).

low repayment capacity) which is critical in credit transactions. Second, borrowers are selected during village meetings, selection is eventually relatively transparent. This may limit risks of corruption, and thus of leakage of the bank's resources. Finally, as many community-based programs, benefits for the community itself can be important. Such programs contribute to empowering communities, particularly in a country like Vietnam which is recovering from centrally planned decision making. They may contribute to creating social capital within the village by intensifying information flows among villagers. It finally helps in mobilizing local forces such as social networks and traditional social structure to participate in the design and implementation of policies which can improve their efficiency.

The community-based targeting system does not perform well. Even though information may be fairly complete in the village and the targeted population well identified, the existing social structure may act against the intended allocation. This is especially true for programs targeting a sub-population, such as the poor, who occupy a weaker social position within villages. In the VBSP lending scheme described above, candidate borrowers first have to convince other villagers in order to get their votes. In contexts where wealth level is strongly correlated with social status and education level, the poor are not well equipped to make their way through such systems. Moreover, the fact that VBSP is highly subsidized reinforces the public good character of its service. Subsidies render loans particularly attractive, and are likely to raise claims of non-targeted villagers over these loans, and community members may not have sufficient power and/or will to deny them access.

Mistargeting in rural credit programs is relatively common in developing countries, even in programs that do not involve community-based targeting. A second explanation to the 'community failure' described above is proposed and explored theoretically by Aubert et al. (2009). The strong correlation between wealth and repayment capacity pushes credit agents whose incentives are drawn on the repayment performance of their clients, to select wealthier household. That such mechanism is at place in the present case is also well conceivable. In the case of VBSP, the 'credit agents' are in fact village MO trained by VBSP to ensure high repayment, and their ability to do so is sanctioned each following year with the delivery (or not) of new loans in the village. Not only village MO but also villagers under this system face the incentive to target households with high repayment capacity and thus the wealthier.

The subsidization of the loans and the participatory nature of the village-based political negotiations of who gets credit and how much is likely to increase credit rationing at the expense of the poor, and to raise uncertainty about credit access, especially among the poorest.

5.4.3 Is there an interaction between the formal and informal sectors?

We now examine the interaction between participation in the formal sector and in the informal sector. A justification for government intervention on rural credit markets is the expansion of formal credit supply in order to reduce dependence of farmers on the informal sector, which is considered insecure and too expensive. In section 5.3, we saw that objectives from this perspective are far from being reached in the area, since the informal sector remains the first lending sector, despite advantageous formal contracts. We saw in addition that many farmers continue to rely on more expensive and risky contracts to finance agricultural input and consumption although they could obtain cheaper formal loans.

As seen in the introduction, this phenomenon is commonly observed in developing countries. Yet reasons behind it differ, and might vary from one context to another. Understanding how both sectors interact provides valuable information on the ability of the formal sector to respond farmers' needs, to substitute for the informal market, and to reach its objective. Whether the persistence of the informal sector has to be imputed to borrowers' preference or to institutional constraints calls for different policy responses. Possible interventions range from adapting formal supply to farmers' needs (i.e. by offering more short term loans, small loans, or saving facilities), to transforming the lending procedure so as to better reach the poor, or to reinforcing the legal system to secure informal transactions (Ray, 1998, chapter 14).

The empirical approach is as follows, we jointly estimate determinants of participation in the formal and informal sector using a bivariate binary estimation approach:

$$\begin{cases} y_i^F = 1\left[\beta^F X_i + u_i^F > 0\right] \\ y_i^I = 1\left[\beta^I X_i + u_i^I > 0\right] \end{cases} \tag{5.3}$$

where y_i^F and y_i^I denote participation of household i in the formal and informal sector in 2007, X_i is a vector of household and lender characteristics, u_i^F and u_i^I are the error terms, which we assume to be jointly normally distributed. The correlation between both error terms accounts for the non-independence between both equations and is written ρ. β^F and β^I are parameters to estimate.

We focus on credit market participation for the year 2007, a year in which our data is the most complete. The semi-formal and the informal sector are grouped into one 'non-formal' sector. Formal loans are all disbursed in cash and are fully fungible, they can theoretically substitute any other loan types from all sectors. If products from both sectors were perfect substitutes, we would find that participation in one sector reduces participation in the other. When both sectors are perfectly independent, participation in one sector does not affect participation in the other. A test on the parameter ρ is thus a test of independence and/or substitutability or complementarity between both sectors.

Variables contained in X_i are described in the appendix of this chapter, Table 5.10. Starting from the same set of explanatory variables explaining demand and access for formal loans above, we proceed to some adjustments. Land area is now the area managed by the household in 2007. We include a dummy variable indicating whether the household had a running debt from one of the formal banks at the beginning of 2007, and a variable measuring households' liquidity position at the beginning of the year. This variable is the sum of 'liquid' assets held by the household in value. An asset is 'liquid' if it can be sold easily and rapidly on the market, this includes both durable goods and livestock. We expect this value to affect participation in one or the other sector, both as a determinant of demand and of access (if being perceived as an index of creditworthiness).

Results are presented in Table 5.7. We report in columns (3) to (6) the marginal effect of each variable in predicting the four joint probabilities (i.e. participation in formal sector only, in both sectors, in the non-formal sector only, and non-participation). Marginal effects express the effect of an incremental increase of a variable from its mean value on each probability. The overall significance of the model is good (the single probit models not reported here also exhibit good significance and goodness-of-fit estimates).

The Wald-test on ρ yields a p-value of 0.055, we reject the hypothesis independence between both equations. This probability to participate in one sector is affected by the participation in the other sector, hence both decisions are not independent.

Among human capital indicators, age of the household head, its gender, the presence of men and women, and the education level of household members influence sectoral choice. The probability of participating in the formal sector increases with age, but decreases above 58 years (87th percentile). It increases for male-headed households, and with the availability of male labor force in the household. The marginal effect however is not significantly different from 0 in predicting the four probabilities, but we find that the likelihood of participating in the formal sector only decreases with the presence of women in the household. The education level of active members reduces the probability of participation in the non-formal sector, but again its marginal effect is different from 0 in none of the four predictions.

Table 5.7: Joint determinants of participation in the formal and non-formal sector, 2007. Bivariate probit estimates and marginal effects

	Formal sector		Non formal sector		Marginal effects predicting							
	P(F=1) (1)		P(NF=1) (2)		P(F=1, NF=0) (3) dydx		P(F=1, NF=1) (4) dydx		P(F=0, NF=1) (5) dydx		P(F=0, NF=0) (6) dydx	
	coef.	**z-stat**	**coef.**	**z-stat**	**x 100**	**z-stat**	**x 100**	**z-stat**	**x 100**	**z-stat**	**x 100**	**z-stat**
Age head	–0.119***	(3.14)	0.049	(1.14)	**0.710**	(2.01)	**–2.327**	(1.72)	–0.822	(0.55)	**2.440**	(2.85)
Age head squared	0.001***	(2.77)	–0.001	(1.48)	**–0.006**	(1.83)	0.019	(1.39)	–0.822	(0.52)	**–0.021**	(2.56)
Male head+	0.693*	(1.85)	0.004	(0.01)	2.775	(1.55)	**23.756**	(1.97)	**–25.763**	(2.12)	–0.768	(0.18)
Education	0.032	(0.73)	–0.112*	(1.77)	0.241	(1.10)	0.282	(0.24)	–1.016	(0.86)	0.493	(0.83)
Active females	–0.132	(0.85)	0.144	(0.72)	**–1.196**	(1.69)	–2.475	(0.59)	5.777	(1.45)	–2.105	(1.14)
Active males	0.285**	(2.08)	–0.093	(0.65)	0.577	(1.10)	5.657	(1.56)	–5.848	(1.62)	–0.386	(0.31)
Dependants	0.114	(1.51)	–0.010	(0.09)	0.563	(1.41)	2.994	(1.56)	**–3.975**	(1.93)	0.417	(0.47)
Paddy	0.145	(1.22)	0.034	(0.55)	0.231	(0.72)	3.954	(1.30)	–3.503	(1.17)	–0.683	(1.05)
Paddy squared	–0.022*	(1.74)	0.000	(0.22)	**–0.052**	(1.73)	**–0.582**	(1.87)	**0.578**	(1.86)	**0.056**	(1.72)
Upland	0.021	(1.43)	0.001	(0.04)	0.048	(0.80)	0.562	(1.37)	–0.548	(1.50)	–0.061	(0.35)
Upland squared	0.000	(0.92)	0.000	(0.29)	0.000	(0.60)	–0.004	(0.99)	0.004	(1.13)	0.001	(0.38)
Land title+	0.169	(0.59)	–0.376	(1.08)	–0.092	(0.05)	1.815	(0.23)	–0.931	(0.11)	–0.792	(0.18)
Participation political	0.096	(0.50)	–0.341	(1.48)	1.270	(1.44)	1.573	(0.30)	–5.408	(1.02)	2.566	(1.29)
Participation non political	0.271	(1.24)	0.476*	(1.68)	–0.275	(0.28)	8.424	(1.42)	–4.843	(0.84)	–3.306	(1.23)
Kinh language	0.010*	(1.82)	–0.002	(0.39)	**0.077**	(2.10)	**0.292**	(1.78)	**–0.462**	(2.86)	0.093	(1.06)
Current formal debt+	–0.476***	(2.58)	0.051	(0.19)	–1.074	(1.18)	**–11.858**	(2.36)	**11.799**	(2.44)	1.133	(0.44)
Elevation	0.001**	(2.15)	–0.001*	(1.81)	**0.012**	(2.97)	**0.039**	(1.97)	**–0.067**	(3.59)	**0.016**	(1.66)
Distance to village centre	–0.455	(1.54)	0.728	(1.52)	**–3.469**	(1.97)	–9.075	(1.19)	**18.056**	(2.26)	–5.511	(1.26)
Distance YC	–0.006	(1.44)	0.005	(0.91)	**–0.041**	(1.97)	–0.146	(1.26)	**0.240**	(2.24)	–0.053	(0.95)
Distance to shop	0.198***	(3.00)	–0.029	(0.44)	**0.606**	(2.05)	**5.292**	(3.05)	**–5.696**	(3.13)	–0.201	(0.32)
Distance to market	–0.007	(0.47)	0.000	(0.00)	–0.016	(0.19)	–0.299	(0.66)	0.260	(0.56)	0.055	(0.24)
Liquidity (log)	0.133**	(2.14)	0.008	(0.13)	**0.682**	(2.03)	**5.903**	(2.39)	**–6.376**	(2.58)	–0.209	(0.32)
Farm income (share)	–0.020***	(3.29)	0.024***	(3.87)	**–0.107**	(3.00)	**–0.284**	(1.83)	**0.561**	(3.19)	**–0.170**	(3.54)
Wage income (share)	–0.009	(1.34)	0.009	(1.23)	–0.039	(1.23)	–0.042	(0.23)	0.163	(0.81)	–0.082	(1.30)
Observation	287				χ^2				326.1***			
Log-likelihood	–215.5				p–value wald test (ρ=0)				0.055			

z-stat are based on robust standard errors – Bold numbers are marginal coefficients significantly different from 0 at the 10% level of error probability – *p>0.1, ** p>0.05, *** p>0.01.

Probability of participation in the formal sector decreases with the area of paddy land managed by the household, once this area reaches 3.3 acres per capita (53rd percentile). Looking at the marginal effects, we also observe a mild negative slope of this variable in explaining the probability of participation in the formal sector (columns (3) and (4)) and a mild positive slope in the probability of borrowing from the informal sector only, and not borrowing at all. As already seen, the ability of households to speak Kinh matters for their access to formal credit. This is observed again, with a significant and negative slope in predicting participation in the nonformal sector only. As expected, households which had a debt running in the formal sector by the beginning of 2007 are less likely to participate in the formal sector. Although in theory, a household is not allowed to hold two debts from a formal lender at the same time (VBARD or VBSP), a number of households do so in practice. We do not find a negative effect on the probability of borrowing from the formal sector only, but a negative effect on the probability of participating in both sectors, and a positive effect on the probability of participation in the informal sector only.

Geographic location explains a substantial share of sectoral choice. All else being equal, the elevation of a household's homestead positively influences participation in the formal sector, but negatively affects participation in the non-formal sector (the coefficient remains small in magnitude). Households living closer to the city center are more likely to participate in the formal sector only, and less likely to borrow from the non-formal sector only. The estimated marginal effects are quite large (an additional traveling minute leads to a decrease in the probability to take part in the formal sector only of 3%, and increase the probability of falling in the third category by 18%) an effect that has to be imputed to transaction costs. The further the village from a shop the more likely its inhabitants are to participate in the formal sector. The relationship is insignificant regarding distance to market. The lack of availability of informal credit sources is a reason for taking part in the formal sector. The variable measuring households' liquidity position positively affects participation in the formal sector, but negatively affects the probability of being a non-formal borrower only. One interpretation is that those in need of liquidities turn to the informal sector rather than to the formal one where the risk of being rejected due to underestimated creditworthiness is higher. Finally, and as observed before, household income sources matter in sectoral choice. Households specialized in farming have a higher probability of participating in the non-formal sector, and a lower probability of taking part in the formal sector.

The results shown here indicate that there is scope for improvement regarding the share of the formal sector in credit transactions in the area. Formal credit substitutes expensive informal loans to some extent, and households accessing formal loans are likely to reduce their participation in the informal sector. This extent is limited however, as we observe a large share of formal clients borrowing additionally from the non-formal sector. In 2007, 94% of VBSP clients and 87.5% of VBARD clients also contracted informal loans. One interpretation is that loan amounts offered by VBSP are insufficient in covering farmers' needs. The average loan size offered by VBSP for a duration of three years is lower than the total average amount borrowed by even poorer farmers for one year. This may however, be untrue in the case of VBARD. A second interpretation is that informal and formal loans are imperfect substitutes, and formal loans are used to finance other purposes than those financed through informal contracts. This also explains the low share of formal loans used in financing agricultural input. As seen in section 5.3.1, farmers have a high demand for credit to finance agricultural input and food consumption. Long- and medium-term credit offered by VBSP and VBARD are unlikely to serve this need.

5.5 Impact of the government micro-credit program on welfare

VBSP bank was established to supply credit to the market segment left uncovered by VBARD, and by doing so to empower farmers in their decision-making and reduce rural poverty. The credits offered by VBSP have been tailored for the poor: interest rates are low, and thanks to low amounts, and high duration default risks are quite low. The previous sections revealed that despite its announced objectives the bank has a low targeting efficiency among the poor. We have also seen that the ability to fully substitute expensive informal loans is limited as farmers' credit demand often exceeds the average loan amounts provided by banks.

The program is still largely subsidized and constitutes an important spending for the government. In this section we explore the impact of the program on the welfare of its participants, namely on households' levels of expenditure per capita, and asset value.

5.5.1 The propensity score matching approach

When the assignment of treatment (here participation in VBSP credit program) is not random, the usual impact estimator (i.e. difference-in-difference estimator) leads to biased estimates, due to selection bias and the presence of confounding factors. We use the propensity score matching (PSM) technique to correct these biases. The main advantage of PSM over other impact methodologies (i.e. instrumental variables or control function methods for instance) is that it does not impose a functional form on the distribution of the outcome variables. A main disadvantage, however, is that estimates are likely to remain biased if selection is too heavily based on unobservable characteristics (Blundell and Dias, 2009). The PSM approach suits well the characteristics of the credit program under study, and the richness of the dataset available enables us to control for a number of observable characteristics. As seen in section 5.4.2 selection of households into the program depend households' demand for credit, household characteristics that are observable by village MO and other villagers, as well as village characteristics influencing resource availability. Each year and in every village, there are eligible households willing to participate in the program who are not able to because of limited resources. Most villages however apply a turnover in allocating credit, and households considered 'eligible' whose access was denied because of limited resources one year, will have a higher probability of receiving a loan in the next years. In such condition, it is reasonable to expect that once controlled for observed village characteristics and household creditworthiness indicators, the allocation of credit is close to random.

The PSM approach was introduced by Rosenbaum and Rubin (1983). It is based on the conditional independence assumption (CIA), i.e. the assumption that conditional on a set of observable characteristics X, the assignment of treatment is independent from the studied outcome.

$$Y_0, Y_1 \perp D \mid X, \qquad \forall \quad X \tag{5.4}$$

where Y_0 and Y_1 are the outcome values of non-treated and treated households, respectively, D denotes treatment and takes the value 1 when the individual has received treatment and 0 when otherwise. X is a vector of observable characteristics. Conditional on X therefore, the assignment of treatment can be considered as random. Matching method thus suggests to match control and treated households on X to obtain an impact estimate. In practice however X is likely to be multidimensional, which renders matching difficult. Rosenbaum and Rubin (1983) go further and demonstrate that if the CIA holds for X, it also holds for the propensity score $P(X) = P(D = 1|X)$ which is the probability that a household receives treatment conditional on X:

$$Y_0, Y_1 \perp D \mid P(X), \qquad \forall \quad X \tag{5.5}$$

Treated and control households are matched within classes of $P(X)$. A further requirement is that the 'common support' condition is satisfied: the propensity score must be estimated only for classes of $P(X)$ in which control and treated are of sufficient number to be compared. This also means that each household included in the impact estimation must have a positive probability to be both treated and untreated. We estimate the impact of the treatment on the treated, noted *ATT* (as opposed to the *ATE* which measure the impact on the overall population) which is estimated as follows:

$$\tau_{ATT} = E_{P(X)}\{E[Y_1|D=1, P(X)] - \hat{E}[Y_0|D=0, P(X)]\} \qquad (5.6)$$

τ_{ATT} is thus simply the difference in outcome between treated households (first term) and the estimated counterfactual $\hat{E}[Y_0|D=0, P(X)]$ that is obtained through observations within the control subsample. The calculation of the counterfactual can be based on different algorithms that assign different weights to control observations in their comparison with the treated observations.

5.5.2 Estimation of the propensity score and matching algorithms

The treatment variable denotes participation of a household in the VBSP credit program in 2005–2006. Outcome variables measure household level of per capita total expenditure, food expenditure and non-food expenditure, which includes clothing, but excludes durables, education and health expenditure. Because these variables were measured over the 12 months preceding the survey, the data overlap both years 2006 and 2007 and will not be affected by household credit access in 2007. Other outcomes relate to the value of durable goods purchased by the household in 2005 and 2006 and households asset value at the beginning of 2007. As seen in section 5.3.2 and 5.4.3 we suspect VBSP credit program to have a higher impact on consumption and asset acquisition than on agricultural outcome. Formal credit accounts only for a minor share in households' agricultural input purchase. Investment in livestock however is one of the most widely cited purposes of borrowing a VBSP loan, and we therefore test for an impact on the value of household livestock holding in 2007.

The selection of explanatory variables used to estimate the propensity score was done in two steps. Starting from the set of variables used in modeling demand and access to VBSP credit in section 5.4.2, we adjust the model to ensure a good matching quality, and avoid an over-parameterization of the propensity score which may jeopardize balancing condition and matching quality Caliendo and Kopeinig (2008). We remove insignificant variables based on the estimates of the standardized bias before and after matching (see Caliendo and Kopeinig

(2008, p. 48)). Estimates of the propensity score matching are presented in the appendix of this chapter, Table 5.11.

In order to increase comparability between treated and control observations, we adjust the estimation sample. In a first stage, we drop observations from households that had no demand for credit during the period considered (this leads to a reduction of the number of observations by 23%). That is, we exclude households that did not apply for credit during that period because of no need (but keep households we were too discouraged to apply). Since results are likely to be under-evaluated by the presence of households that have received in that period a VBARD loan, we thus exclude in a second stage all households that have been VBARD clients between 2004 and 2006 (column (2)).

Two different algorithms are employed as a check for robustness. A one-to-one approach (the nearest neighbor method) is inappropriate here considering the relatively small sample size. The so-called radius matching compares the treated households with all control households whose distance in propensity score does not exceed a certain distance (caliper). The second method is the kernel matching method. Each control household is compared with the treated one, but the weights are assigned to each matching partner according to distance between treated unit and the control ones.

Statistics on the matching quality of both algorithms and within both subsamples are shown in Table 5.8. In all cases, results indicate a good matching quality. Standardized bias is reduced by more than 90% in most cases. The Pseudo-R^2 after matching is also significantly reduced, and covariates which were jointly significant before matching become insignificant after matching.

5.5.3 Impact estimation

The simple comparison of outcomes (*t*-test) between treated and control households shows important and significant differences. Treated households have higher level of food and non-food per capita expenditure, and a higher level of total expenditure per capita. They possess more durables and a higher value of livestock. After matching however, these differences vanish, and we do not observe any significant impact within the subpopulation of households having a positive demand for credit of VBSP credit program.

Turning now to the second subsample of non-VBARD clients, we find, as expected, that the estimated differences after matching are larger. The estimates indicate no impact on household per capita expenditure, but significant differences in asset indexes. Results indicate that households purchased more assets during

Table 5.8: Impact of VBSP credit program, results from propensity score matching estimates

	Mean treated	Mean control	Before matching Diff.	Before matching t-stat	After matching Radius (caliper=0.09) Diff.	Radius t-stat	After matching Kernel (bandwidth=0.07) Diff.	Kernel t-stat
A. All household with positive demand for credit								
Daily per capita expenditure	17.45	15.17	**2.52**	(2.52)	0.12	(0.11)	0.03	(0.03)
Daily per capita food expenditure	9.72	8.87	**0.96**	(1.74)	–0.11	(0.17)	–0.16	(0.25)
Daily per capita non food expenditure	4.07	3.07	**1.09**	(2.64)	0.27	(0.53)	0.24	(0.45)
Value of durables purchased between 2005–2006	5,243.7	3,769.7	1,339.9	(1.54)	1,301.9	(1.25)	1,394.7	(1.31)
Value of durables begining of 2007	13,174.9	6,155.8	**7,332.6**	(2.65)	1,208.3	(1.11)	1,137.7	(1.03)
Value of livestock held by household in 2007	17,555.4	12,603.2	**4,786.1**	(2.81)	–721.8	(0.35)	–1,161.4	(0.55)
N Treated				61		56		56
N Control				162		162		162
Standardized bias				23.98		2.00		2.81
Pseudo-R^2				0.20		0.01		0.02
P-value χ^2				0.000		1.000		1.000
B. Non VBARD clients and positive demand for credit								
Daily per capita expenditure	15.64	13.97	**2.56**	(2.21)	–0.92	(0.67)	–0.99	(0.72)
Daily per capita food expenditure	8.65	8.13	0.90	(1.45)	–0.92	(1.28)	–0.99	(1.35)
Daily per capita non food expenditure	3.62	2.86	**1.12**	(2.26)	0.05	(0.08)	–0.01	(0.02)
Value of durables purchased between 2005–2006	5,115.9	3,201.9	**1,837.8**	(1.80)	**2,748.1**	(2.13)	**2,781.7**	(2.11)
Value of durables beginning of 2007	8,213.5	5,205.7	**3,811.0**	(3.66)	**3,151.1**	(2.34)	**3,131.9**	(2.29)
Value of livestock held by household in 2007	15,893.5	11,346.8	**5,916.8**	(2.78)	–694.9	(0.26)	–572.3	(0.21)
N Treated				44		41		41
N Control				99		99		99
Standardized bias				1.88		2.05		2.56
Pseudo-R^2				0.25		0.02		0.01
P-value χ^2				0.000		1.000		1.000

All figures are in thousand VND, and estimated in constant 2007 prices, using Consumer Price Index published by General Statistical Office of Vietnam – In 2007, 16.000 VND ≃ 1US$ – Bold figures indicate significance at the 10% level of error probability the time that they were clients, resulting in a higher total value of durables owned at the beginning of 2007. The estimated impact on durable purchase is substantial and average 2.75 million VND ($\simeq$ 175 US$). The estimated difference in durables holding amounts up to 3.1 million ($\simeq$ 194 US$). As a matter of comparison, the average annual income in the area amounts to 25 million VND ($\simeq$ 1562 US$), and the estimated differences in durables holding represent about 12.4% of this income. VBSP loan amounts average 7.5 million VND, we find that on average, about 40% of this amount is spent on the purchase of durable goods. If we do not observe impact on expenditure in the short run, in the long run the impact might become positive as the accumulation of durables will contribute to increase households' risk bearing capacity and their ability to undertake long term investments (Zeller and Sharma, 2000). Our estimations indicate furthermore that this impact is likely to increase as the microcredit program targets households having no access to VBARD loans[45].

Finally, we find no impact on livestock holdings. This is surprising at first, as a large share of VBSP loans are officially spent on financing investment in livestock (i.e. referring to declared loan uses of borrowers). Semi-structured interviews, conducted by Kim Anh (2010) and one author of this paper, reveal that livestock investment is indeed the best and most frequent purpose declared when applying for a VBSP loan. After receiving the loans however, many farmers use the money to finance durables or housing expenses. A respondent cited by Kim Anh (2010) explains that some farmers avoid sanctions from the bank by borrowing one of their neighbor's buffalo the day the bank staff comes to check investment. Estimates presented here confirm that such practice is more than anecdotal. The low level of risk associated with VBSP loans on the one hand, and the fact that many wealthy and credit unconstrained households benefit from these loans on the other hand, explains this low impact.

---

45 We check for the robustness of estimated impact using the Rosenbaum bounds methodology described in Rosenbaum (2002) and DiPrete and Gangl (2004). Results are not presented here but are available upon request. They indicate that the estimated impact on durable purchase in 2005–2006 is sensitive to the presence of unobservable factors explaining in the selection into treatment, thus the estimated impact is likely to vanish as the weight of unobservable variables increases in the selection process. However the estimated impact on asset holding in 2007 is relatively robust to the presence of unobservable characteristics. The estimated impact stays significant for values of gamma ranging from 1 to 1.6.

5.6 Discussion and policy recommendations

Three striking results emerged from the analyses. First, the ability of formal lenders to respond to local demand is limited, and consequently, the informal sector remains the first lending sector. Overall however, results show that demand for credit in the area is very high, and the market is rather fluid, since most farmers, even the poor, are able to access credit from various sources to finance consumption and the purchase of agricultural inputs. Despite advantageous contracts, the formal sector remains a secondary credit source. Our results suggest that if high transaction costs and the important risks associated with VBARD loans can shift preferences of some farmers towards other lenders, such explanations do not hold true in the case of the VBSP, the state-owned microcredit bank. VBSP operates inside villages and offers cheaper and safer loans than any other lender. The choice to borrow from the informal sector instead of VBSP is rather induced by the limited credit supply, and thus by a rationing of clients. The shortcomings of the microcredit program lie not only in the limited supply but also in some of its settings. One major drawback is that the program does not present itself as a reliable banking partner. Given the politicized and complex lending procedure of the bank involving decision-makers at the commune and village levels, as well as mass organizations, it is almost impossible for a household to predict the possibility of accessing a VBSP loan. Villages may receive a great deal of credit in one year, and too little or no credit at all in other years. Rather than a banking partner, the bank simply acts as a government institution implementing credit distribution policies which are riddled with administrative and political uncertainties. This micro-credit program therefore does not respond to the high need for regular credits to finance consumption and agricultural input. Long repayment period prevents clients from contracting any other formal loans for three years. Yet, the amounts lent by VBSP are too low to cover households' total credit demand. Under such conditions, VBSP program is unlikely to substitute informal loans, but may in fact increase farmers' reliance on informal lenders in the years following the contract.

Beyond preference and credit rationing, we find evidence that the persistence of the informal sector is explained by an imperfect substitutability between formal and informal credit. Thanks to good level of trust, and good information access, the informal sector is able to provide flexible loan contract terms with variable maturation at moderate interest rates, and most transactions are in-kind. Comparatively, the 'one size fits all' approach of the VBSP, combined with its lack of predictability is not competitive. Informal transactions in addition entertain social relationships within communities and other social networks. We also find that many farmers seek to diversify their loan portfolio, presumably as a way to reduce both default risks and their dependence on a single sector.

Second, our results show that Vietnam's intervention through a 'commercial' bank and a 'policy' bank for the poor does not induce a more equitable allocation of credit than the existing informal sector. Initially designed for the poor, VBSP loans rarely reach their targets. The community-based targeting system, while offering some serious conceptual advantages, is not functioning well and seems particularly inadequate in the context of a microcredit program. Allocation results in a distribution that is worse for the poor than a random allocation would be, the system instead favors the wealthier and those having access to other sources of formal credit. Decentralized and community-based development has gained much interest among policy makers and development practitioners in the last ten years, yet some authors have expressed their doubts about the benefits to be expected from such an approach Conning and Kevane (2002); Platteau and Abraham (2002). The case presented here provides a good illustration of the inefficiencies and mistargeting that may arise in such programs. 'Endogenous community imperfections', as labeled by Platteau and Abraham (2002) on the one hand, and incentives of credit agents (here village MO) being heavily drawn on the repayment performance of clients on the other hand are two factors hindering accurate targeting Platteau and Abraham (2002); Aubert et al. (2009). The mistargeting constitutes an important leakage of state resources which urgently needs to be addressed.

Finally, agricultural productivity is affected only indirectly by the government's rural finance policy. VBARD appears to serve the non-farm sector rather than financing agricultural investments. The microcredit program also contributes little to investment in the off-farm and the livestock sectors. The low level of risk associated with these loans, and the fact that they are borrowed by well-off households, partly explains this low impact. As a consequence, the program does not contribute to increasing households' per capita expenditure, at least not in the short term. This last finding is in line with existing empirical evidence showing little or no impact of microfinance on welfare in the short term (cf. Banerjee et al. (2010)). Furthermore, by setting the interest rate below inflation level, the credit program discourages saving and encourages low return investments, such as the purchase of durables for consumptive purposes. While long term positive income effects can be expected, the extent of these effects on poverty reduction will be limited as long as the richer households capture these resources.

Several policy adjustments need to be undertaken in order to address these inefficiencies. We suggest some recommendations which could help with improving both the efficiency and equity outcomes of Vietnam's policy. First, restoring the financial viability of the VBSP is important to ensuring its sustainability, and freeing the state from unnecessary expenditure. One way to do this is for VBSP to expand its activities to collecting deposits and thereby to mobilize capital. Given the high demand for liquidity in the area, such a service is expected to meet a high demand under the condition that the product is attractive to farmers and that it is well advertised. Contrary to general wisdom, the poor have a demand for saving products and do save already through informal and sometimes insecure means (Zeller and Sharma, 2000; Rutherford, 2000).

The other, more imminent way to achieve financial sustainability is for VBSP to increase its lending rates. In addition to increasing revenues, such a measure may raise borrowers' incentives to invest in profitable activities, in the off-farm or livestock sector, and increase targeting efficiency as incentives to exclude the poor at the benefit of the wealthy and politically connected people are reduced. Our data show that households' demand for credit, even by the poor, is relatively inelastic. Using the data published on the MIX market database (MIX, 2011) and calculations proposed by (Armendariz de Aghion and Morduch, 2007, p. 237) we estimate that lifting the interest rate per annum up to 14.3% would enable the bank to become financially sustainable[46]. Thanks to low lending costs permitted by the reliance on village organization, the bank is able to recover its sustainability while proposing loans at rates remaining significantly lower than those found in the informal sector and just above the inflation rate level.

Second, more effort must be spent on targeting the poor. In addition to better training of bank staff and village MOs, reducing the selection power of the latter is necessary. The establishment of a monitoring mechanisms or incentive-based systems ensuring that selected clients are those targeted could avoid important mistargetings. Some mechanisms are in place already, documents stipulating household income level and signed by the village head must be submitted to the bank for approval. However, as observed in the fields, many borrowers under estimate their revenue on these documents[47]. The application of sanctions to

46 This figure assumes a default rate of 5%.

47 In some cases we could verify directly, as the data set collected also contains income and poverty classification data.

punish such practices may help preventing them. In addition, a better targeting would reduce the bank's clientele and enable the VBSP to increase loan size and frequency, and thereby improve the predictability of obtaining a loan from the view point of borrowers.

Finally, in Vietnam as in many developing countries and given the specific demand for short term credit, the scope for substituting informal sector through formal contracts is limited. The informal sector thanks to good information access and high levels of trust is able to offer more flexibility in in contracts and is likely to continue playing a major role in the area. Policies may seek to reinforce the efficiency of this sector instead of weaken it. Policies strengthening the legal system so as to enlarge enforcement capacities of informal lenders would contribute to secure financial transactions in this sector, raise its efficiency and lower interest rates.

5.7 Conclusion

Vietnam, with the establishment of two state-owned banks and a highly subsidized microcredit program is applying old recipes to the rural finance sector. Using a rich dataset on the credit transactions and access from 300 farm households randomly selected in the district, we have explored the rural credit market and its actors, investigated econometrically the determinants of demand and participation in the two formal banks, as well as the interaction of this market sector with the other lenders that are active in the area. Finally, we assessed the impact of the government's microcredit program on household welfare. Analyses conducted have uncovered a number of inefficiencies in terms of financial sustainability, poverty outreach and impact that need to be addressed in order to save state resources which could be spent more effectively to reduce poverty. We formulated above some recommendations which could partly solve these inefficiencies such as: expanding bank's activities towards collecting saving, identifying better targeting mechanisms for the state microfinance program to concentrate on the poorest segment of the population, raising formal interest rates in this program, strengthening the legal system so as to offer informal lenders better enforcement capacities.

The present study does not show the long term impact of the micro-credit program since its implementation. It is in fact likely that the program has helped many households to improve their living standards and has contributed to reducing poverty in the study area. This result, as shown, was achieved at a high cost. VBSP, the state microcredit program, given the high level of subsidies received, resembles more of a low-cost transfer program rather than a competitive micro-credit program. Such a model heavily draws on Vietnamese rich web of political and social organizations and is hardly replicable elsewhere. Its sustainability and persistence in the long run fully depends on government's will and capacity to maintain the program. One of the initial objectives of the microfinance movement from the earliest stage of its development, was to increase poor clients' creditworthiness over time so as to link them with existing formal credit markets. This objective is clearly not pursued by VBSP but yet could enable the bank to target populations in need while empowering former clients.

Beyond the case study, findings in this paper carry out general lessons for financial market development in developing countries. First, this case study shows that accordingly with theory, under good information availability and a high level of trust, the informal sector can efficiently provide credit to all households at relatively moderate interest rates. Policies enhancing levels of social capital, information flows and reinforcing legal institutions can contribute to reducing transaction costs and thereby improving the efficiency and outreach of the informal sector, and enable it to cover a market segment that cannot be reached by formal lenders. Second, our results highlight the pitfalls of community-based targeting systems especially as applied to credit projects, as well as the inefficiencies of 'supply-driven' credit program (as opposed to demand-driven) which results almost inevitably in credit rationing and misallocation. The study finally highlights that contract diversity and reliability of credit access in particular are pivotal feature in rural credit programs to respond farmers' credit demand, and enhance investment capacity.

Acknowledgement – The paper benefited greatly from discussions with Le Thi Ai Van, Dinh Thi Tuyet Van, Aliou Diagne, Alwin Keil, Florence Milan, Emily McNulty and colleagues from the Uplands Program. We are thankful for the precious collaboration of Le Thi Ai Van and Dinh Thi Tuyet Van in the data collection. This research would not have been possible without the support of Prof. Pham Thi My Dung from the Hanoi University of Agriculture, the Uplands Program, the local branches of VBSP and VBARD and of the Women Union in Yen Chau. We are thankful to all village heads and respondents for their time and willingness to participate in the survey. This research was funded by the Deutsche Forschungsgemeinschaft (DFG), German research fundation, through the special research program SFB 564. All errors remain ours.

5.8 Appendix

Table 5.9: Principal component factors of the wealth index, descriptive statistics

Variable	Mean	S.D.	Factor loadings
Cooking meal in separate kitchen+	0.75	(0.43)	0.650
Cooking meals in one of the rooms+	0.22	(0.41)	–0.632
No toilet+	0.1	(0.30)	–0.706
Lighting source is Elecricity+	0.82	(0.39)	0.735
Lighting source is Gas/Oil/Kerosene lamp+	0.09	(0.29)	–0.715
Roof made of big tiles+	0.29	(0.46)	–0.646
Roof made of tiles+	0.62	(0.49)	0.678
Floor is made of earth+	0.15	(0.35)	–0.584
Nb years household was classified as poor in 2002–2006	1.06	(1.87)	–0.753
Value of cupboard*	1 546	(1 875)	0.763
Value of living room set*	284	(468)	0.612
Value of electric fans*	73	(92)	0.716
Value of video player*	327	(363)	0.597
Value of color television*	1 286	(1 210)	0.723
Eigen value			6.503
% Variance explained			0.465
Overall KMO			0.869

+ indicate dummy variables (1=yes, 0=no)
* Value in 2006, means displayed in thousand dong, variables are logged in the factor estimation

Table 5.10: Description of household and village level explanatory variables (alphabetical order)

Variable	Description	Mean	S.D.	Min	Max
Active females	Number of female actives	1.34	0.70	0	4
Active male	Number of male actives	1.23	0.77	0	6
Age head	Age of household head	43.72	12.42	17	88
Dependants	Number of dependants	2.35	1.56	0	12
Distance to market	Distance to closest market (in km)	6.19	7.19	0	27
Distance to shop	Distance to closest shop (in km)	0.32	1.29	0	6
Distance village centre	Distance between household homestead and its village centre (in km)	0.34	0.35	0.03	3.30
Distance Yen Chau	Distance to Yen Chau city (traveling minutes by motorbike)	43.37	36.64	5	160
Durables 2004	Value of durables owned by the household in 2004 (in million VND per capita)	2.00	5.65	0	85.97
Education	Average education level of adults	4.90	2.91	0	12
Elevation	Elevation of homestead (meter above the sea level)	515.98	240.07	281	1088
Farm income	Percentage of household cash income derived from farming in 2002	85.75	25.30	0	100
Formal debt	Household had a debt running in the formal sector at the beginning of 2007+	0.45	0.50	0	1
Income shock	Household has experienced an income shock in the five years preceding the survey (complete crop failure, loss of a working animal, loss of a working adult, payment of dowry)+	0.74	0.44	0	1
Kinh language	Percentage of adults speaking Kinh language	85.07	21.42	0	100
Land title	Household holds a land title+	0.87	0.34	0	1
Liquidity	Value of 'liquid' assets owned by households in 2007 (durables and livestock) (Million VND)	12.13	9.81	0	66.46
Male head	Household head is male+	0.91	0.28	0	1
Paddy area 2004	Area of paddy managed in 2004 (in acre per capita)	3.58	4.57	0	59.95
Paddy area 2007	Area of paddy managed in 2007 (in acre per capita)	3.77	4.69	0	59.95
Participation - P	Weighted mean of household member participation in political organizations[a]	0.25	0.41	0	3.25
Participation - NFP	Weighted mean of household member participation in non political and financial organizations[a]	0.26	0.44	0	1
Poor classification	Household was classified as poor at least once in 2004–2006+	0.26	0.44	0	1
Poorer village	Household lives in a relatively poorer village	0.20	0.40	0	1
Richer village	Household lives in a relatively richer village	0.30	0.46	0	1
Upland area 2004	Area of upland managed in 2004 (in acre per capita)	25.05	18.02	0	110.67
Upland area 2007	Area of upland managed in 2007 (in acre per capita)	29.61	19.63	0	123.33
Wage income	Percentage of household cash income derived from wage in 2002	10.81	21.55	0	100
Zone I	belongs to Zone I, well-off economic zone	0.41	0.19	0	1
Zone III	belongs to Zone III, poor economic zone	0.40	0.31	0	1

+ indicates dummy variables (1=yes, 0=no) In 2007, 1 Million VND $\simeq$ 62.5US$

[a] Political organizations are those which directly depend on the state control: communist party and mass organizations (Farmer, Women, Youth, Veterans, Elderly and Fatherland front Unions). Non political and financial organizatiosn regroup all other kind of associations, excluding credit groups. Weights account for the degree of participation of each household member, with the highest weight assigned to group leaders, and a lower weight assigned to non active members.

Table 5.11: Estimation of the propensity score matching – Probit estimates

	Participation in VBSP credit program (2005–2006)			
	(1)		**(2)**	
	Coef.	**z-stat**	**Coef.**	**z-stat**
Education	0.085*	(1.86)	0.064	(1.10)
Active males	0.514***	(3.30)	0.493**	(2.39)
Active females	–0.232	(1.35)	–0.061	(0.27)
Paddy area	–0.128*	(1.81)	–0.082	(0.94)
Paddy area squared	0.005	(1.53)	0.003	(0.71)
Upland area	0.024	(1.31)	0.048**	(2.04)
Upland area squared	0.000	(0.94)	–0.000*	(1.77)
Durables 2004	0.011	(0.48)	–0.005	(0.06)
Classified as poor	–0.436	(1.52)	–0.562*	(1.68)
Farm income share	–0.914	(1.18)	–1.314	(1.17)
Wage income share	–1.990**	(2.07)	–2.877**	(2.12)
Distance village centre	–0.636*	(1.70)	–0.402	(1.00)
Richer village	0.260	(1.08)	0.283	(0.84)
Poorer village	0.706*	(1.91)	0.460	(1.04)
Zone I	–0.055	(0.22)	–0.113	(0.33)
Zone III	–1.275**	(2.35)	–1.446**	(2.50)
Constant	–0.369	(0.45)	–0.156	(0.14)
Observations		223		143
Pseudo-R^2		0.196		0.248
χ^2		51.3***		43.8***
% correct predictions		0.76		0.74

* p>0.1, ** p>0.05, *** p>0.01

6. Ethnic heterogeneity and the formation of social capital in rural communities of Northern Vietnam

Abstract

Using data from rural communities in which ethnic heterogeneity was induced within through involuntary resettlements policies in the 1960s, we estimate the exogenous effect of ethnic heterogeneity on individual participation in local organizations and households' social network capital. The effect on participation depends on organizations' political nature and the public nature of managed goods. We find no direct impact on social network capital but an indirect effect through interactions with identity and participation. Results do not confirm theoretical predictions of a negative relationship but show that ethnic heterogeneity can encourage bridging connections, and, as such, foster innovation and economic development.

6.1 Introduction

Over the past decade, rural communities have been recognized increasingly more by scholars and practitioners as essential actors in economic development, and have been given greater influence over the design and implementation of development policies and projects. The so-called decentralized and community-driven development projects represent, according to Mansuri and Rao (2004), about US$7 billion of the World Bank's portfolio. In Vietnam as well, after de-collectivization in 1988, communities have been entitled to a central role in the implementation of anti-poverty policies, infrastructure development projects, and other reforms such as land allocation. This trend is based on the belief that communities, thanks to better information access and higher enforcement capacities, have a comparative advantage, when it comes to allocating resources and implementing projects equitably and efficiently, over remote national decision-makers.

Despite this strong interest, concerns have been raised. Some argue that communities within and across countries are not uniform in their ability to implement projects effectively; they point to the existence of community-level features that can greatly influence the economic performance and the success of the decentralized approaches (see for instance Platteau and Abraham (2002); Conning and Kevane (2002); Khwaja (2009)). Social capital, broadly defined as the "norms and networks that enable people to cooperate" has appeared as a powerful concept in explaining the economic performance of communities, and their ability to implement such projects successfully (Woolcock, 1998; Khwaja, 2009).

Since the emergence of the concept in economic literature, and the seminal work of Coleman (1990) and Putnam (1993), the literature has grown exponentially study the economic importance of social interactions. At the micro-economic level, evidence shows that social capital, measured by participation of community members in local organizations or social networks, has a positive impact on income levels (Narayan and Pritchett, 1999; Haddad and Maluccio, 2003; Grootaert and Narayan, 2004), on credit access (Okten and Osili, 2004), and on the adoption of new technologies (Isham, 2002)[48]. Khwaja (2009) finds the impact of community social features on the success of infrastructure projects in Northern Pakistan to be significant, i.e. projects are more likely to succeed in "good" communities; but also finds that these effects can be altered by project design, and thus good projects can also succeed in "bad" communities.

Surprisingly, little research has focused on assessing the determinants of the formation of social capital, i.e. what makes a community "good" or "bad" in terms of participation in collective life, cooperation, or formation of social networks. Alesina and La Ferrara (2000), Glaeser et al. (2002) and La Ferrara (2002) studied the determinants of individual participation in local organizations in the U.S. and in Tanzania. A larger set of the literature has examined the determinants of social network formation in rural contexts (Udry and Conley, 2004; De Weerdt, 2004; Goldstein et al., 2005; Fafchamps and Gubert, 2007; Santos and Barrett, 2010; Chantarat and Barrett, 2011). Many of these studies have focused on the roles of identity, geography and other socio-economic characteristics in network formation, but none – to the best of our knowledge – have yet looked at community features as factors that either inhibit or enhance the formation of social networks.

Ethnicity is often indicated as an important element of social capital formation. Ethnic heterogeneity, in particular, is frequently seen as one factor that inhibits interactions among individuals in a social group because of preference, communication barriers, and social sanction effects (Glaeser, 2001; Bates and Yachovlev, 2002; Alesina and La Ferrara, 2005; Putnam, 2007). The issue, however, has rarely been investigated empirically in micro-economic studies. Alesina and La Ferrara (2000) study the effect of heterogeneity on participation

48 See Durlauf and Fafchamps (2004) for a more complete review.

in U.S. communities and find it to be negative. Looking at the effect of heterogeneity on collective action in Nepalese forest communities, Varughese and Ostrom (2001) find no apparent relation. La Ferrara (2002) explores the determinants of participation in rural communities of Tanzania and finds that tribal fragmentation does affect participation incentives, albeit in an ambiguous way, depending on the type and entry rule of the social group considered. In many areas, identification and interpretation of ethnic heterogeneity effects is particularly difficult because communities were formed endogenously. Voluntary migration from "bad" to "better" locations (with respect to market links, social conditions, or areas of targeted policy intervention, such as food aid) causes heterogeneous communities to often be better located, or to differ from other communities in further variables of social cohesion.

The study takes place in a mountainous district of Northern Vietnam where we know that ethnic heterogeneity was induced in the 1960s at the village level by involuntary, government organized migration. Vietnam is an ethnically diverse country: 54 ethnic groups are officially recognized by the government. The Kinh majority represents 82% of the population, and the remaining 18% is made up of the other 53 ethnic minority groups. These groups vary both in size and in their degree of integration with the Kinh majority. Ethnic minorities live mostly in mountainous areas, in the Northern Uplands where this study takes place, and the Central Highlands regions. On average they are poorer than the Kinh majority, due to greater remoteness, lower endowments in natural- and other types of capital, as well as lower returns on characteristics (van de Walle and Gunewardena, 2001; Baulch et al., 2007; Epprecht et al., 2011). The relationship between the state and the ethnic minority groups has evolved dramatically over time and remains, to this day, complex (Michaud, 2009). Resettlement of Kinh households in mountainous areas, and the settlement of swidden cultivator tribes[49], was required after independence in order to, among other objectives, integrate ethnic minorities into the country's political project.

Our data was collected in 2007 in a mountainous district of the Northern Uplands where (re)settlement policies have been implemented after independence from France (1954). The area is now occupied both by ethnically homogeneous communities in which migration has seldom taken place, and heterogeneous communities in which involuntary migration organized by the government has altered village composition. We show that the degree of ethnic heterogeneity

49 Swidden cultivator tribes are groups traditionally practising slash-and-burn cultivation with seminomadic settlement patterns, such as the H'Mong.

within communities is strongly affected by these (re)settlement policies, but systematically associated with better locations and endowment, indicating a relatively exogenous process of population mixing. This situation therefore makes for a unique opportunity to empirically study the role of ethnic diversity.

This article seeks to contribute to the empirical literature about the economic effects of ethnic heterogeneity, making use of this distinct setting to analyze ethnic heterogeneity within villages exogenously, and therefore to test for its impact. We explore the effect of heterogeneity on aspects of social capital which have shown to greatly matter in economic development, namely participation in local organizations and households' social network capital. Furthermore, we also develop what we believe is a novel measure of financial social network capital based on a measure of households' borrowing capacity (credit limit) from their social networks relative to their income.

Results point to an ambiguous effect of ethnic heterogeneity. First its effect on participation appears to depend on the type of organization considered, its political affiliation, the entry rule or whether it manages public or club goods. Second, the results show no direct effect of ethnic heterogeneity on households' social network capital, but an indirect effect through interactions with other variables.

The rest of the paper is organized as follows: we review the literature in Section 6.2 and present the settings and the data in Section 6.3. The econometric models are presented in Section 6.4, results are presented and discussed in Section 6.5 and we conclude in Section 6.6.

6.2 Ethnic heterogeneity and social capital, a review

6.2.1 Defining social capital

This section presents the concept of social capital and introduces the theoretical literature that is motivates the empirical work in the later sections.

The literature exploring the effect of social capital has grown exponentially since the emergence of the concept in the late 1990s. Despite the success it has had in scientific literature, defining and delimiting this concept 20 years after its emergence, as something tangible and measurable, remains a challenging issue (Dasgupta, 2000; Schuller et al., 2000; Durlauf and Fafchamps, 2004; Hayami, 2009). Putnam (1995, p. 664) famously defined social capital as the "features of social life – networks, norms and trust – that enable participants to act together more effectively to pursue shared objectives", a definition which has influenced much of the research in the last two decades. Earlier works, by Bourdieu (1986) and Coleman (1990), presented individuals as the owners and first beneficiaries of social capital. Bourdieu (1986, p. 286) defined social capital as "the aggregate of the actual or potential resources which are linked to possession of a durable network". Social capital thus encompasses structural (networks, associations) and cognitive dimensions (norms, trust) and is both individual and communal in dimension. Without entering the discussion of whether it is best seen as an individual asset or a community feature, we note that the study of the determinants of social capital requires an individual perspective. The decisions to join a network, to participate in collective organizations or to cooperate in the elaboration of public goods are indeed made individually (Sobel, 2002; Glaeser et al., 2002; Mogues and Carter, 2005). The 'amount' of social capital available at the individual or community level however, has implications for both individuals and communities due to the existence of important spillover effects, which are difficult to disentangle.

Theoretical works and empirical evidence suggest that the level of social capital in a community is at least partly endogenously determined, and that individuals decide whether to join a group and cooperate, based on economic, social or institutional conditions (Woolcock and Narayan, 2000). Glaeser et al. (2002) and Mogues and Carter (2005) have developed economic models of investment in social capital, in which individuals decide to invest based on their objective utility function. Alesina and La Ferrara (2000),Glaeser et al. (2002) and La Ferrara (2002) have explored empirically the determinants of participation in local organizations in Tanzanian and U.S. communities based on similar hypotheses. Community features such as income inequality, ethnic diversity and remoteness are all potentially important factors. The literature on collective action also highlights the institutional framework, in addition to exogenous community characteristics, such as the definition of property rights and the existence of clear cooperation rules, as important determinants of collective action, and thus of cooperation among individuals (Ostrom, 2000). Here, we focus on ethnic heterogeneity as an influential factor of social capital formation.

Two sets of the literature provide theoretical grounds on which to study the effect of ethnic heterogeneity on social capital in its broader definition. The first relates to a growing segment of literature studying the effect of ethnic heterogeneity on public good formation (social capital being considered as a public good), and the second refers to the literature linking identity to social network formation.

6.2.2 Ethnic heterogeneity and public goods

The literature on ethnic heterogeneity and public good formation usually concludes in a negative relationship (Alesina et al., 1999; Costa and Kahn, 2003; Miguel and Gugerty, 2005; Alesina and La Ferrara, 2005; Habyarimana et al., 2007). The explanations provided, however, have diverged. Some have referred to the lack of social sanction mechanisms available in ethnically and culturally diverse societies. The existence of sanction mechanisms is expected to enhance cooperation among individuals by enhancing one's anticipation of others' behaviours, and securing cooperative choices. This mechanism is invoked by Miguel and Gugerty (2005) to explain the lower level of public spending on schools in rural Kenya in ethnically diverse communities.

Others have referred to preference-based mechanisms which can be divided into two subsets (Alesina et al., 1999; Habyarimana et al., 2007). First, people of different cultural and ethnic backgrounds are expected to have different preferences, and the weak 'commonality of tastes' – and thus of interests in public goods – in ethnically diverse communities impedes agreements on the elaboration of public goods. Second, individuals may have preferences not only on the nature of public goods, but also bout interacting with people of different cultural backgrounds. The latter was proposed by Alesina and La Ferrara (2000) to interpret their results of the negative impact of heterogeneity on participation in local associations in U.S. communities.

Finally a third mechanism, that Habyarimana et al. (2007) refer to as a 'technical mechanism' relates to cultural material that homogeneous communities can rely on when designing public goods and that enables them to cooperate more effectively. This material includes language, the ability to understand anticipate each other's behaviour, and the existence of rules facilitating cooperation. Other studies suggest, however, that the relationship may not always operate in the above-mentioned direction, i.e. that more homogeneity is conducive to more effective cooperation. Alesina and La Ferrara (2005) refer for instance to a set of the literature showing that heterogeneous groups, by bringing together people with different skills and backgrounds, may be better equipped to solve problems in an innovative manner. This third mechanism, however, refers to the 'impact' of social capital (i.e. to the outcome of cooperation) rather than to its determinants, although community members may choose (not) to participate in collective actions based on their anticipation of the high (low) returns of their interactions with others.

Based on these theoretical considerations, ethnic heterogeneity is mostly expected to lower levels of social capital, i.e. of cooperation among individuals.

Empirical evidence, as pointed out in the introduction, remains relatively scarce and seems to vary from one cultural-geographic context to another.

6.2.3 Identity and social network formation

A second portion of the literature that sheds light on the role of ethnic heterogeneity on social capital looks at the role of identity in the formation of social networks. The economic importance of identity, in its broad definition, is explored by Akerlof and Kranton (2000)[50]. The literature on endogenous social network formation also highlights the importance of identity. Fafchamps and Gubert (2007) explore the formation of risk sharing networks in the Philippines and find that geographic proximity, which may be related to kinship, is the main factor of association, while economic factors do not appear to be significant. Goldstein et al. (2005) study factors of inclusion in mutual insurance networks in Ghana by looking at the probability of which one obtains help within and outside his household when a shock occurs. They find that membership in the village's main lineage, and participation in secular organizations are the two main factors that increase the probability of receiving help outside of the household. Santos and Barrett (2010) investigate the formation of information networks in Ghana to assess the relative importance of identity and material interest (measured by differentials in productivity). They conclude that both identity and interest matter in information networks. The role of identity in network formation appears, therefore, to depend heavily on the type of goods exchanged through networks, whether information, credit or insurance. Ethnic heterogeneity, due to a higher dispersion of identities and kinship within groups, is likely to hamper social networks formation.

Akerlof and Kranton (2000) defined identity as "a person's sense of self", suggesting that elements that form one's identity vary between social environments and from one person to another, and thus that identity is a relative and rather subjective concept. Following this approach, identity is defined not only by a person's type (whether male or female, from a certain ethnic group or

50 Unlike these authors, our approach of identity does not encompass an economic dimension such as income. Since the focus of this work is on ethnicity, we concentrate on exogenous and immutable aspects of identity, such as sex, ethnicity, family name, place of birth, etc.

born in a particular place) but also by the identity of the other group members, and her differences from others in the same group. The subjective identity of an individual living in a community of her own ethnicity differs from what her subjective identity would be if she lived in a community of a different ethnic group, and differs again from what it would be if the community were very mixed or international.

This subjective identity is likely to affect both her willingness and capacity to connect with others. Chantarat and Barrett (2011) develop a theoretical model on social network formation showing how individuals decide to connect with others based on the cost of establishing a relationship. This cost is a function of the social distance, thus the difference in identity between two individuals. One can consider in this framework that heterogeneity affects these costs of establishing relationships. Two individuals that have fairly different identities may feel distant in a homogeneous environment, but this subjective distance may shrink in a more heterogeneous community.

To sum up this literature review, the causal effect of ethnic heterogeneity on social capital formation is likely to be important. In a collective view of social capital, theoretical evidence points to a negative effect on both cooperation and on the level of participation in collective action. Empirical evidence, however, has not always confirmed this relationship, and some research suggests that the institutional framework matters more. With the more individual perspective of social capital, or social network capital view, heterogeneity has an ambiguous impact. On the one hand, identity is shown as an important element of social network formation – because individuals that have close identities are more likely to connect – and thus the dispersion of identities induced by ethnic heterogeneity is likely to impede social network formation. Ethnic heterogeneity on the other hand, by affecting the definition of subjective identities, as perceived by individuals themselves or by their neighbours, is likely to have an indirect and ambiguous effect on social network formation.

6.3 The data

6.3.1 The settings

Vietnam is an ethnically diverse country. According to the official classification, the Kinh majority (82% of the country's population) shares the nation with 53 other ethnic groups, so-called ethnic minorities. The empirical study takes place in the Northern Uplands, one of the most ethnically diverse regions of Vietnam. The area nowadays is home to 31 ethnic groups, and observed levels of diversity are high even within communes, because of the complex settlement patterns of populations over time in the region (Michaud et al., 2002). After independence from France in 1954, the regime made it a priority to unify the people, and to integrate ethnic minorities into the nation's political project, as well as to foster development of upland areas and ensure food security in the lowlands. The resettlement of lowland villages into the uplands, and the settlement of swidden cultivator communities, was undertaken in this perspective, and often accompanied by acculturation programs (Hardy, 2000; Pelley, 2002; McElwee, 2004; Baulch et al., 2007; Michaud, 2009; Friederichsen and Neef, 2010).

Our study area is Yen Chau, a rural district located along the national road N6, which links Hanoi with the provinces of Son La and Dien Bien Phu. Three ethnic groups inhabit the area: the Thai which account for 55% of the district's population, the H'mong (20%), and the Kinh (13%). Other ethnic groups (the Sinh Mun and Kho Mu) are also present but represent only a small fraction of our sample (<1%). The Thai were the first settlers and have occupied the lowlands mainly, while the H'mong have settled primarily in the uplands (Neef et al., 2003). (Re)settlement of villages in the study area began in the early 1960s, a few years after the country gained independence from France. In some cases, entire villages were resettled by the government, as is true for eight villages in our sample. Additionally, Kinh households were moved from the Red River Delta to Yen Chau and resettled into existing villages. Host villages were selected based on land quality and availability. Newly settled households received compensation in the form of land, housing assistance or monetary support. Host villagers in most cases were not compensated. Seven out of our 20 sample villages were asked to host new comers between 1960 and 1982. The number of households resettled varied between 10 and 100 households, which in some cases doubled the initial village population.

Ethnic groups present in the area have different customs, social organizations, languages, clothing and eating habits. The Thai, for instance, are traditionally organized into hierarchical social structures with matrilineal lineage (Mellac, 2006), while in the H'mong case, social structure is more horizontal with patrilineal lineage system (Corlin, 2004). Customs regarding, for example, the establishment of newly married households, the inheritance of land, or land allocation among community members differs between the three groups. Traditional practices however have been strongly attenuated by several decades of integration policies, resettlement interventions and land reform (McElwee, 2004; Friederichsen and Neef, 2010).

Villages, although not officially recognized as an administrative unit, are the places where major decisions affecting households' livelihood are made. The allocation of untitled land, the granting of formal credit, the definition of households' poverty status entitling them to participate in specific assistance programs, and access to irrigation are, among many others, important decisions that are made at the village level. The literature further emphasizes the strength of social relations within villages, and the political challenges that they constitute for the state. An old saying in Vietnam stipulates that the "King's law gives way to the village custom" (*Phep vua thua lang* in Vietnamese). A recent example is provided by Kerkvliet (2005) the author shows how villagers in the lowlands opposed the state administration during decollectivization through 'everyday politics', and how these contestations and local adjustments led to redefinitions of the national policy. In the uplands, the implementation of the 1993 land law in particular, aimed at individualizing land access, has resulted in tensions and opposition from upland communities (Bryant, 1998; Sikor, 2004)[51].

6.3.2 The survey

The data were collected in 2007 and 2008 in Yen Chau district. A representative sample of 300 households was selected following a two-stage cluster sampling procedure. A village-level sampling frame was constructed encompassing all villages of the district[52], including information on the number of resident households. First, 20 villages were randomly selected using the Probability Proportionate to Size (PPS) method. Next, 15 households were randomly selected in each of the selected villages using updated village-level household lists. Since the PPS method accounts for differences in the number of resident households between villages in the first stage, this sampling procedure results in a self-weighing sample.

51 Wirth et al. (2004); Sikor (2004); Mellac (2006) provide specific examples in Thai communities, and Corlin (2004) discusses this issue in the H'mong context.

52 The urban centre and four communes located along the border with Laos for which research permits were difficult to obtain at the time of interview were excluded from the sampling frame.

The survey covered a wide range of topics and included a detailed module on social capital, as well as on households' credit experience and access to different sources. The social capital module gathered detailed information on individual participation in organizations, on levels of trust, reliance on networks in case of shocks, etc. Additionally, a census of all residents[53] was conducted in each of the 20 sample villages, by which we obtained information on the ethnicity and family names of all residents. The settlement history of each village was retraced through focus group discussions with the village elders and structured interviews with knowledgeable villagers.

6.3.3 Descriptive statistics

Table 6.1 presents some descriptive statistics of village and household characteristics in homogeneous and heterogeneous villages. Ethnic and name fragmentation are measured with a Herfindahl-Hirschman fragmentation index:

$$FRAG_{\upsilon} = 1 - \sum_{i=1}^{T} \pi_{i\upsilon}^{2} \tag{6.1}$$

$\pi_{i\upsilon}$ is the share of households belonging to a group i in village υ; T is the total number of groups in the village, defined by ethnicity or family names. The index $FRAG_{\upsilon}$ can be interpreted as the probability that two individuals taken randomly in the village belong to different ethnic groups, or carry different family names. Of the 20 sample villages, nine are completely homogeneous, and 11 are ethnically mixed. The degree of heterogeneity within those mixed villages varies widely, and the ethnic fragmentation ranges from 0.01 to 0.5 in these villages. The observed fragmentation score for names varies from 0.27 to 0.82 in the most diverse villages.

53 The literature suggests that clanship used to be important within the Thai and H'Mong ethnic societies (Mellac, 2006; Corlin, 2004) but that through efforts of acculturation and integration, and the successive land reforms the importance of clans nowadays is nowadays limited. Family names are the closest indicators of clanship that we can use. However, we must note that two individuals carrying the same family name do not necessarily belong to the same clan, but the probability that they do is higher. We recorded 31 different family names for a total of 2,296 households within twenty villages.

The strong correlation between the degree of heterogeneity and the occurrence of resettlement in the village, as well as the number of households that have been resettled indicates that these resettlements constitute the major source of observed heterogeneity within villages. We find no association between heterogeneity and geographic factors such as elevation or distance to markets, to the city centre or to paved roads. Nor do we find any significant association between heterogeneity and income or land inequality (measured with coefficients of variation). Kinh households are present only in heterogeneous villages, while the presence of Thai and H'Mong does not vary between village types.

Looking now at the second part of the table displaying household characteristics, we find significant associations between heterogeneity and the level of household per capita expenditure, but no apparent association with households' income diversification. The size of individual farms is also larger and positively correlated with heterogeneity. This corroborates interview findings indicating that households were resettled in villages where land was available. When asked about other selection criteria, none of the villages in which households were resettled could respond, and the lack of correlation with other village characteristics shown above also confirms this finding.

We find significant associations between heterogeneity and migration. A large share of household heads was born outside the province and migrated to the village at some point, as is also true for the older generations. Finally, participation in local organizations does not differ on average between villages. We find, however, that homogeneous villagers participate more in mass organizations while heterogeneous villagers participate more in civic organizations.

Mass organizations (or unions) are associations created by the state at the time of independence as a way to convey and apply at the local level the principles and policies of the communist governance. They are present at each administrative level, from the state to the village. At the local level, each of the six unions (cf. Footnote in table 6.1) carries out different tasks and objectives. All play an important role in a village's everyday life as village union leaders sit on the village board and participate in village level decision making. The women's, farmer's, veteran's and youth unions act as credit agents for the state microfinance bank (the Vietnam Bank for Social Policies), and organize the allocation of Bank's loans among villagers. The farmer union in some villages is responsible for contract arrangements between villagers and input and output traders and for organizing agricultural extension and therefore plays an important role in farming decisions. Having one member taking part in these unions is therefore important for accessing public resources and information. In some villages, participation is even

Table 6.1: Village and household-level characteristics by village ethnic heterogeneity

	Homogeneous (n=9) Mean	(S.D.)	Heterogeneous (n=11) Mean	(S.D.)	T-test/ χ^2 test p-value	Spearman correlation[e] p-value
A. Village characteristics						
Ethnic fragmentation index	0	(0)	0.17	(0.17)	0.009	
Name fragmentation index	0.53	(0.15)	0.69	(0.11)	0.013	0.011
Resettlement took place in that village+	0.11	(0.33)	0.64	(0.50)	0.017	0.061
No. of resettled households	1.11	(3.33)	33.54	(43.51)	0.031	0.064
Elevation (m.a.s.l.)	501	(246)	527	(250)	0.820	0.890
Distance to						
- market (km)	3.7	(4.9)	8.7	(8.9)	0.147	0.265
- city center (travelling minutes)	40.6	(33.8)	45.5	(40.9)	0.777	0.546
- paved road (walking minutes)	18.3	(19.0)	13.6	(20.6)	0.607	0.369
Within-village coefficient of variation						
- per capita daily expenditure	1.86	(0.73)	2.12	(0.76)	0.445	0.271
- titled upland	1400	(1115)	1501	(897)	0.826	0.529
- titled paddy fields	461	(1027)	352	(291)	0.739	0.051
Population (inhabitants)	463	(277)	579	(219)	0.312	0.426
Share of households from ethnic						
- Kinh	0.01	(0.01)	0.19	(0.31)	0.097	0.002
- Thai	0.77	(0.44)	0.72	(0.37)	0.772	0.084
- H'Mong	0.22	(0.43)	0.08	(0.27)	0.405	0.906
B. Household characteristics	**(n=135)**		**(n=165)**			
Per capita daily expenditure	14.5	(6.5)	16.3	(6.5)	0.017	0.013
Share of off-farm income in total cash income	0.03	(0.14)	0.04	(0.14)	0.584	0.557
Farm size (m^2 per capita)	3,185	(2,026)	3,667	(2,322)	0.020	0.059
HH head born outside the province+	0	(0.00)	0.14	(0.35)	0.000	0.000
HH head's father born outside the district+	0.07	(0.09)	0.19	(0.39)	0.000	0.000
Participation in local organizations						
- All	1.37	(0.62)	1.36	(0.65)	0.862	0.949
- Mass organization[a]	1.05	(0.47)	0.94	(0.38)	0.027	0.226
- Communist party	0.03	(0.11)	0.05	(0.14)	0.311	0.975
- Professional organization[b]	0.01	(0.06)	0.04	(0.17)	0.107	0.795
- Civic organization[c]	0.13	(0.21)	0.19	(0.30)	0.046	0.278
Credit limit from friends and relatives per unit of income[d]	400	(501)	624	(1,053)	0.024	0.087

[a] Farmer Union, Women Union, Youth Union, Eldery Union, Veteran Union and Fatherland front Union.
[b] Extension clubs and other professional associations.
[c] School committee, Health committee, Parent group, Civil security, Environmental group.
[d] In Thousand VND. In 2007, 16.000 VND $\simeq$ 1US$.
[e] Spearman correlation test between variable and ethnic fragementation index.
+ indicates dummy variables

compulsory, and individuals are de facto considered members of their corresponding union (i.e. Women's union for women, Farmer's union for farmers, etc.). This explains, together with the high interests these unions represent, the high observed rates of participation in these organizations.

Civic associations also manage public resources encompassing school committees, health committees, and security groups. However, entrance into these associations is not compulsory, but results more of a free choice. We also notice the low participation rates in the communist party in both village types. Whereas in urban and lowland areas, participation in the communist party may provide some advantages, the role of the party in rural areas is more limited, even though the village communist party secretary also sits in the village board. Finally, we report, in the last line of Table 6.1, the measure of household informal (from friends and relatives) credit limit per unit of income, which we use later on as an indicator of households' social network capital (see Section 6.4.2). This measure of credit access draws on the credit limit approach developed by Diagne et al. (2000). The credit limit is a measure of the maximum amount a household could borrow at the time of survey from a certain lender given its current debt to the lender. We focus here on households' credit limits from their friends and relatives. As explained by Diagne et al. (2000), it is reasonable to expect farmers to have accurate expectations towards their borrowing capacity thanks to their extensive experience in borrowing from different sources due to the seasonality of income earnings. In the study area we find that a household on average contracts six loans per year, of varying amounts and from different sources. About two thirds of these loans are for small amounts (below 200 thousand VND $\simeq$ 12.5 US$ in 2007), and obtained from informal sources.

Table 6.2 reports descriptive statistics on households' credit limits from different sources by terciles of income (proxied by the level of per capital daily expenditure). We find for both the formal and the informal sector that the richest 33% of respondents have a credit limit three times greater than that of the poorest third of respondents. This gap is further accentuated for credit from friends and relatives. On average, the richest third can borrow four times more from their social network than the poorest third, indicating the importance of reciprocity in social relationships related to financial issues. The amount one is able to borrow from others depends on how much one is able to lend. We report statistics denoting the share of this credit limit that was being borrowed at the time of the interview. Poor households use a larger share of their credit limit than their richer counterparts, which confirms that this source is particularly important for the poor. The last line in Table 6.1 show a positive association between this measure of social network capital, and the degree of heterogeneity in the village.

Table 6.2: Descriptive statistics on household credit limit, per expenditure tercile

	Poorest (n=100)		Middle (n=100)		Wealthiest (n=100)	
	Mean	**(S.D.)**	**Mean**	**(S.D.)**	**Mean**	**(S.D.)**
Daily expenditure per capita	8.7	(2.3)	14.8*	1.7	23.0*	(4.2)
Credit limit from						
–Formal sector	8,587	(7,467)	19,558*	(23,344)	29,053*	(31,843)
–Semi-formal sector	1,895	(3,430)	2,068	(2,933)	5,388	(20,376)
–Private lenders	8,511	(8,775)	11,937*	(14,129)	20,496*	(37,175)
–Friends and relatives	3,231	(4,754)	9,510*	(17,435)	12,748*	(17,972)
% currently borrowed	20.3	(30.3)	17.9	(29.1)	10.2*	(20.3)
per unit of income	374	(527)	629*	(1,116)	566*	(809)

All figures are in thousand VND. In 2007, 16.000 VND 1 US$.
* Indicate a difference from the first tercile significant at the 5% level of error probability

6.4 Empirical strategy

6.4.1 The participation model

The model of participation we estimate follows closely those of Alesina and La Ferrara (2000) and La Ferrara (2002). It considers that an individual decides to participate in an organization if he derives a positive net utility. The net utility he gains from participating in one type of organization is expected to depend on his own characteristics, the characteristics and identity of the household to which he belongs, and his village's attributes:

$$B^k_{ij\upsilon} = \alpha^k X_{1ij\upsilon} + \beta^k X_{2j\upsilon} + \gamma^k I_{j\upsilon} + \delta^k D_v + \epsilon_{ij\upsilon} \tag{6.2}$$

where $B^{k*}_{ij\upsilon}$ is the expected utility of an individual belonging to household j in village i, when participating in association υ; $X_{1ij\upsilon}$ is a vector of individual characteristics; $X_{2j\upsilon}$ is a vector of household characteristics; $I_{j\upsilon}$ denotes the identity of the household; D_υ are village attributes; α^k, β^k, γ^k and δ^k are parameters to be estimated and $\varepsilon_{ij\upsilon}$ is the error term. Letting $P^k_{ij\upsilon}$ denote the observed participation of the individual i in association k, we expect that the individual decides to take part in the association k if his or her net utility is positive:

$$\begin{aligned} p^k_{ij\upsilon} &= 1 \text{ if } B^{k*}_{ij\upsilon} > 0 \\ p^k_{ij\upsilon} &= 0 \text{ if } B^{k*}_{ij\upsilon} \leq 0 \end{aligned} \tag{6.3}$$

Explanatory variables included in the participation model are described in the Appendix, Table 6.5. Individual-level variables account for age, gender, education, and occupation. We include a variable indicating whether the member speaks more than one language, as this helps to reduce social distance between two individuals of different ethnic groups.

The included household-level variables account for the effect of households' demography, wealth level and identity. Demography is expected to ambiguously affect an individual's decision to participate, depending on the labour constraint faced by the household and the nature of benefits gained through membership. Variables account for household size and the presence of dependants, which affect labour allocation within the household and determine the time available for participation. One would expect members of larger households to participate less in associations for which benefits linked to membership are less than proportional to the number of members. On the other hand, when the labour constraint is binding, the effect of household size becomes positive.

We control for households' wealth-level by using a composite index of 14 variables expressing different dimensions of long-term wealth (see the principal component factors of the wealth index already presented in the appendix of chapter 5, Table 5.9). We prefer this index of long-term wealth to a measure of per capita expenditure which was measured at the same time as participation, and is more volatile, and probably endogenous.

Finally, we include variables denoting the households' identity. The variables take into account three dimensions of identity: ethnicity, household name and the places of birth of the household head and spouse. The first two dimensions are captured through ethnic dummy variables and an index of inherited social distance (ISD) between the household and its neighbours according to ethnicity and family name. We finally include a dummy variable indicating whether the household spouse or head are native to the village[54].

The ISD variable captures the relative aspect of one's identity (see Section 6.2.3), i.e. the difference of the household from its neighbours along variables denoting immutable identity features. A commonly used approach to calculate such an index is the Euclidean distance. Here, as we measure the distance between a household and all other villagers, we could write:

$$ISD_{jv} = \sqrt{\sum_k (x_{jv}^k - \overline{x}_v^k)^2} \tag{6.4}$$

54 We have seen earlier that among the three ethnic groups present in the area, two have patrilineal traditions while the Thai follow matrilineal structures, and traditions regarding the establishment of newly married couples also differ between these groups. We thus account for the place of birth of both the spouse and the head.

where $ISD_{j\upsilon}$ is an index of inherited social distance of household j in village υ; $k = [1, 2]$ refers to the type of characteristics, i.e. ethnicity and name; $x^k_{j\upsilon}$ is the observed characteristic k of household j in village υ; and $\overline{x}^k_{\upsilon}$ is the average of x^k in village υ. Because ethnicity and lineage are discrete variables, we convert them into dummy variables and rewrite the inherited social distance index as such:

$$ISD_{j\upsilon} = \sqrt{\sum_k (1 - \overline{x}'^k_{j\upsilon})^2} \tag{6.5}$$

where $\overline{x}'^k_{j\upsilon}$ is now the share of villagers sharing the same characteristics k as the household j. $ISD_{j\upsilon}$ thus increases with the number of villagers being dissimilar to household j. Finally, we include village-level control variables that capture geographic characteristics and village composition, among which is the index of ethnic heterogeneity described in Section 6.3.3.

We run separate individual-level regressions to estimate the determinants of participation in each organization type as a function of the individual, household and village characteristics described above and using probit estimates, with sandwich estimators of the standard error. We test for the endogeneity of the ethnic fragmentation variable using an instrumental variable approach. The instrument used is a dummy variable indicating whether resettlement policy took place in the village after the independence. Results are displayed in Table 6.3 and discussed in Section 6.5.1.

6.4.2 The social network capital model

We now turn to the second model that we build to estimate determinants of households' social network capital. We focus here only on the financial dimension of social network capital, through a measure based on households' access to credit from their friends and relatives (see section 6.3.3). It takes into account one aspect of individual social networks that matters greatly in rural development. Social networks represent an important source of credit, particularly in areas where the credit market is missing or fails to provide loan to all households at market price. Extended family and intracommunity networks, thanks to better information access and to better monitoring and enforcement capacities, are able lend to their networks at low interest rate, making this source particularly attractive, especially for poor households. Because of the covariant nature of credit demand and shocks in rural areas, this source of credit cannot fully respond to households' demand for capital, but strongly enhances farmers' risk bearing and consumption smoothing capacities (Zeller and Sharma, 2000).

Table 6.3: Probit and IV Probit estimates of the determinants of participation in local organizations

	All organization (1) Probit		Mass organization (2) Probit		Communist party (3) Probit		Professional organisation (4)[a] IVProbit		Civic organization (5)[b] IVProbit	
	dF/dx	**t-stat**	**dF/dx**	**t-stat**	**dF/dx**	**t-stat**	**dF/dx**	**t-stat**	**dF/dx**	**t-stat**
Age	1.071***	(2.79)	0.796*	(1.85)	0.705***	(3.40)	1.432***	(3.53)	1.302**	(2.07)
Age squared	–0.009**	(2.19)	–0.005	(1.14)	–0.006***	(2.84)	–0.015***	(3.33)	–0.018**	(2.27)
Male+	–5.391**	(2.34)	–7.465***	(3.09)	0.903	(0.86)	–1.722	(1.21)	5.453***	(2.67)
Education	1.670***	(4.02)	1.010**	(2.27)	1.687***	(4.94)	0.586**	(2.27)	0.806**	(2.16)
Speaks two languages+	3.612	(0.94)	6.469	(1.58)	6.986*	(1.80)	1.105	(0.76)	3.809	(0.97)
Farmer+	9.212***	(2.99)	15.618***	(5.07)	–1.054	(0.97)	–4.050***	(3.61)	0.593	(0.21)
Head or spouse+	4.301	(1.32)	6.149*	(1.81)	–0.788	(0.48)	–3.180**	(2.22)	10.011***	(2.67)
Household size	–2.110***	(3.02)	–2.245***	(3.24)	0.208	(0.90)	–0.148	(0.45)	–1.439**	(2.11)
Dependency ratio	22.031***	(3.34)	17.763***	(2.60)	0.407	(0.14)	5.983**	(2.39)	10.229*	(1.88)
Wealth index	1.127	(0.90)	0.094	(0.07)	2.104***	(2.64)	1.646***	(2.60)	2.416*	(1.94)
Inherited social distance	–3.862	(0.62)	0.378	(0.06)	–3.997	(1.59)	2.879	(1.02)	–19.937***	(3.34)
Native+	6.346*	(1.80)	5.842	(1.55)	1.824	(0.84)	–0.996	(0.68)	3.932	(0.97)
Ethnic Kinh+a	–9.078	(1.35)	–8.544	(1.21)	5.696	(1.40)	3.405	(1.32)	3.460	(0.56)
Ethnic H'Mong+a	–26.550***	(3.49)	–33.008***	(4.08)	14.473***	(4.16)	–		3.037	(0.44)
Elevation	0.035***	(3.49)	0.040***	(3.73)	–0.005	(1.25)	–0.002	(0.63)	0.009	(1.16)
Population	–0.010	(1.57)	–0.013**	(1.98)	0.006**	(2.33)	0.003	(1.00)	0.004	(0.83)
Distance to paved road	0.219***	(2.71)	0.221***	(2.62)	0.093**	(2.30)	–0.095*	(1.70)	0.019	(0.23)
Distance to market	0.068	(0.29)	0.086	(0.33)	0.287**	(2.38)	0.036	(0.41)	–0.302	(1.44)
Ethnic fragmentation	0.103	(1.04)	0.117	(1.10)	–0.196**	(2.32)	–0.319*	(1.76)	0.481**	(2.06)
N	986		986		986		986		986	
χ^2	132.2***		148.1***		114.5***		147.1***		136.0***	
P-value Wald-test exogeneity	0.644		0.548		0.548		0.012		0.100	
% participants correctly predicted (0.25 cut-off)	100		100		64.4		65.0		41.6	

Marginal effects reported in percent – *t* statistics in parentheses based on standard errors clustered at the household level

* $p<0.10$, ** $p<0.05$, *** $p<0.01$

+ indicates dummy variables, (yes=1, no=0)

[a] The missing group is Thai. We do not include the 'other ethnic' group dummy (0.1% of the sample) as this leads to the eviction of part of the observations. In column 4, the dummy ethnic H'Mong is also removed as none of the H'Mong households in our sample takes part in such organization and the inclusion of the variable leads to the eviction of a substantial share of the population.

[b] Instrumented variable is ethnic fragmentation, instrumental equation is shown in the appendix.

The dependent variable Y_{jv} is the household's credit limit from friends and relatives per unit of income described in section 6.3.3. It measures how much one is able to borrow from his friends and relatives per unit of income. We argue that this variable is a good measure of household social network capital. First, it relates directly to Bourdieu's definition of social capital (see Section 6.2.1. Second, it is strongly correlated with three important dimensions of a social network: its size; the intensity of the ties linking the household with network's members, in particular their level of trust of the household[55]; and finally the wealth level of members in this network, and their lending capacity.

Based on the literature on social network formation (Granovetter, 1973; Dasgupta, 2003; De Weerdt, 2004; Fafchamps and Gubert, 2007; Santos and Barrett, 2010; Chantarat and Barrett, 2011), we develop an econometric model to estimate the determinants of households' financial social network capital. The model can be seen as a model of investment in social network, where a household's stock is a function of innate and immutable characteristics such as identity, and of social decisions. We use a log-linear specification to estimate parameters of the following equation[56].:

$$\text{In} Y_{jv} = \alpha + \beta X_{jv} + \gamma E_{jv} + \delta P_{jv} + \eta I_{jv} + \lambda D_v + \upsilon_{jv} \quad (6.6)$$

where X_{jv} are household demographic variables, E_{jv} indicate a household's level of human capital, P_{jv} indicate the participation of household members in associations, I_{jv} denotes the identity of the household; D_v are village attributes.

Table 6.5 in the Appendix, section 6.7 describes and summarizes variables included as explanatory variables in the model. In addition to variables controlling the age and education level of adults, and the gender of all members, we include a variable measuring farmers' wealth level, using the same asset-based index as in the previous model which may be perceived by others as a measure of households' creditworthiness. As in the previous model, we account for three dimensions of identity: ethnicity, inherited social distance and birth place of the head and the spouse. The variable for participation takes into account participation in all types of organizations.

---

55 Most loans from friends and relatives were borrowed in 2007 at 0 interest rate (70%), and on average, interest rate was significantly lower all other lenders, for equivalent amounts and maturation.

56 A non-negligible share of households (6%) declared having no access to credit from their friends and relatives, and a log-transformation of the dependent variable leads to the eviction of these observations. We replace the dependent variable as '0' for these observations. Our results are robust to different treatments of these values, and the log-linear model performs better than the linear one.

We estimate household-level OLS regression on households' social network capital, using village-level controls (described in Table 6.1) in a first stage and village dummy variables in a second stage. As in the previous model, we test for endogeneity of the ethnic fragmentation variable using the resettlement policy as the instrument. We test for hypotheses discussed in section 6.2.3, relative to the effect of ethnic heterogeneity on subjective identity and social decisions by interacting ethnic heterogeneity variable with identity and participation variables Results are presented in Table 6.4 and discussed in Section 6.5.2.

6.5 Results

6.5.1 The determinants of participation

Probit estimates of the participation model are reported in Table 6.3. The first column presents the model that predicts participation in any association, and the following columns present estimates of models predicting participation in different types of organizations, namely mass organizations, the communist party, professional, and civic organizations.

We test for the endogeneity of ethnic fragmentation using an instrumental variable approach. The instrumental equation is presented in Table 6.6 of the Appendix, and shows that the occurrence of involuntary resettlements in the village, the instrument, strongly influences the village ethnic composition and the observed degree of heterogeneity. The results of Wald-tests of exogeneity are presented at the end of Table 6.3, for each equation. The tests lead us to reject endogeneity in the first three models (any organization, mass organization and communist party), but not in the last two models that predict participation in professional and civic organizations. We report in the first three columns results from the simple probit estimates, while in the last two columns, the IV probit estimates are reported (i.e. after instrumentation)[57].

Individual characteristics such as age and gender significantly explain participation. The probability of participation increases with an individual's age, but starts to decrease once the person reaches 64.5 years. Women are significantly more likely to take part in any organization. Individuals with a higher level of education are also more likely to participate, as are those whose main occupation is farming.

---

57 Due to space constraint we do not present estimation results before instrumentation for these two models. Those are available upon request to the authors.

Table 6.4: OLS estimates of household social network capital

DV: log of households' credit limit from friends and relatives per unit of income						
	(1)		(2)		(3)	
	Coef.	**t-stat**	**Coef.**	**t-stat**	**Coef.**	**t-stat**
Household characteristics						
Age adults	–0.021**	(1.98)	–0.023**	(2.22)	–0.025**	(2.37)
Share female	0.174	(0.36)	–0.004	(0.01)	–0.003	(0.01)
Education adults	0.046	(1.11)	0.032	(0.75)	0.024	(0.58)
Farm size	0.009*	(1.70)	0.012**	(2.13)	0.012**	(2.23)
Wealth index	0.267*	(1.73)	0.187	(1.15)	0.142	(0.85)
Inherited social distance (ISD)	–0.015	(0.03)	–0.073	(0.14)	–0.442	(0.75)
Native+	0.393	(1.38)	0.591*	(1.86)	0.891**	(2.59)
Participation	0.208*	(1.78)	0.245**	(2.11)	0.162	(1.10)
Ethnic Kinh+	0.900**	(2.05)	1.367*	(1.79)	0.474	(0.49)
Ethnic H'Mong+	–0.982*	(1.69)	–1.965***	(2.67)	–1.343	(1.46)
Other ethnic+	–1.454*	(1.90)	–1.473**	(2.17)	–1.507**	(2.34)
Village characteristics						
Elevation	0.001	(0.93)	–		–	
Population	–0.000	(0.47)	–		–	
Distance to paved road	–0.000	(0.04)	–		–	
Distance to market	–0.011	(0.61)	–		–	
Ethnic fragmentation	–0.721	(1.08)	–		–	
Village dummies	NO		YES		YES	
Interaction terms						
Ethnic fragmentation x ISD	–		–		5.540	(1.21)
Ethnic fragmentation x Native+	–		–		–3.617*	(1.69)
Ethnic fragmentation x Participation	–		–		1.015	(1.56)
N	299		299		299	
Adjusted-R^2	0.174		0.195		0.206	
F	5.15***		3.40***		3.34***	

t-statistics in parentheses are based on standard errors
* p<0.10, **p<0.05, ***p<0.01

Households with many members and those with a high dependency ratio are less likely to participate, indicating that labour allocation within the household does plays a role. These results however, do not point to a binding labour constraint but shows that the benefits from participation are generated at the household level rather than at the individual level, and thus the marginal net utility from having one more member participating in the organization is negative. The availability of dependants, like elderly members in the household, may give adult members more time to invest in these organizations. In this general model, we do not find that households' wealth level is an explaining factor, which rejects the hypothesis of a budget constraint. Household members whose head or spouse are native to the village have a higher probability of participating, and members of H'Mong households, everything else being equal, are less likely to take part in any organizations. Other identity indexes are found to be insignificant.

Finally, ethnic fragmentation is found to have no direct impact on participation. Household members living in more remote locations (higher elevation or further from a paved road) are more likely to participate. This may indicate that access to services (extension or credit) in the more remote villages is likely to be facilitated through participation in organizations. It may also indicate higher social pressure for villages to participate. Estimates of the first model, as can be seen from the following column are heavily influenced by mass organizations which constitute an important share of households' total participation. Entrance into these organizations is not always a free choice, and indeed determinants of participation in other organizations differ significantly.

In the communist party model, we find that education and the ability to speak at least two languages are significant and positively influencing factors. Household demographic variables do not appear as significant factors, but the wealth level, as measured by the asset-based index, does influence participation positively. The H'Mong, all else being equal, participate more than other ethnic groups. This ethnic group is culturally and politically more marginalized than other ethnic groups in the area, and Vietnam in general (Corlin, 2004). Taking part in the communist party might provide a crucial opportunity for H'Mong people to gain political power. The larger and the more remote the village is, the more likely an individual is to take part in the party.

Members of professional organizations are younger than those of the mass organizations. The probability of participation increases with age, but decreases beyond the age of 49. Non-farmers and members other than heads or spouses are more likely to participate. Participation increases with the dependency ratio and household's wealth level. The probability of participation increases as the distance between the village and the paved road increases, and we find a negative and significant effect of ethnic heterogeneity.

Civic organizations, as explained in section 6.3.3, manage public resources, such as schools and health committees, and environmental or security groups. Unlike mass organizations however, these associations are not state-created, but defined within communities. Induction into the groups is voluntary, but because important public resources are managed by these organizations, membership is expected to be influenced at the village level. The probability of participation increases with age, but decreases after the age of 37. Males and educated individuals, as well as household heads and spouses, are more likely to take part in these organizations. We find, as for mass organizations, that participation decreases with household size, but increases with the household dependency ratio. The asset-based wealth index, which might be related to social status, is found to have a positive impact on participation decision. If ethnicity or the birth place of household heads and spouses are not found to have a significant effects, we find that the index of inherited social distance (ISD), which captures the difference of identity between the household and its neighbours, negatively affects participation. All else held equal, the less similar the household is to its neighbours, the less likely the household members are to enter such organizations. Finally, we find that ethnic heterogeneity positively influences participation, while other village attributes are found to have no impact.

We thus find ethnic heterogeneity to have an ambiguous effect on participation incentives, an effect that depends on the type of organization considered, the entry rule, and on the types of goods that are managed by the organization. Participation in the communist, party, which may be viewed as compliance with the political regime, is negatively influenced. Participation in professional organizations is also negatively affected, while participation in civic organizations is positively influenced. Inherited social distance however, is found to have a negative impact on participation in this last model. We thus do not find ethnic heterogeneity to negatively affect participation in the management of public resources, as is suggested, for instance, by Alesina and La Ferrara (2005). The authors develop a model that shows how ethnic heterogeneity might favour the emergence of private goods, while homogeneity is expected to enhance the formation of public goods. While estimates here do not provide information on the quantity and quality of public goods generated at the village level, they do show that if the quality of public goods were negatively affected by ethnic heterogeneity, as is suggested in other literature, this would not be a result of lower preference mechanisms (and lower participation), but rather of cooperation failures (among participants). The negative sign of inherited social distance shows however, that the competition for managing these public resources might be important, and those who inherited "better" social positions are more likely to participate in the controlling of these goods. The negative indicator for professional organizations is more puzzling, as these organizations manage what can be qualified as club goods. This sign can be interpreted as a greater ability and willingness of individuals living in an homogeneous environment to organize professionally, as a way to increase productivity.

This may occur more in homogeneous environments, where individuals can make use of a common cultural material (see Section 6.2.2) and cooperate effectively.

6.5.2 The determinants of households' social network capital

OLS estimates of the financial social network capital model described in Section 6.4.2 are reported in Table 6.4.

In the first column, we test the direct impact of ethnic heterogeneity on households' network size using control variables to capture village heterogeneity. The small difference observed between Columns 1 and 2 in coefficients and in the adjusted-R^2 indicate that the control variables capture village heterogeneity well. We test, as in the previous model, for endogeneity of the ethnic fragmentation variable using the same instrument as above, i.e. the occurrence of involuntary resettlements of households in a village after independence. With a partial-R^2 of 0.10, an F statistic of 31.4, and a Wu-Hausman test yielding a p-value of 0.39 we confidently treat the variable of ethnic fragmentation as exogenous.

We find that a household's informal lending network size decreases with the age of adult members, but is not affected by the gender of household members nor their education level. Unsurprisingly, households' farm size and wealth level positively influence their borrowing capacity from their friends and relatives. Inherited social distance and whether the household heads or spouses are native to the village are not significant, but once village fixed effects are controlled for, we find that this latter variable has a positive and significant effect. The participation of household members also positively influence households' borrowing capacity from their social networks and this effect is even stronger once we control for village fixed effects. Comparatively to Thai, Kinh households have a larger borrowing capacity from their networks, while the H'Mong and other ethnic groups can borrow less. Once we control for village fixed effects (Column 2), these effects become larger. Ethnic dummy variables in the village fixed-effect model capture ethnicity effects in heterogeneous villages, because ethnicity in homogeneous villages is captured by the village dummies. Therefore, these results indicate that a H'Mong household living in a H'Mong village has higher borrowing capacity than it would in a heterogeneous village.

Village-level variables, including ethnic heterogeneity are found to have no direct effect on households' level of social network capital.

In the third column, we test for interaction effects between ethnic heterogeneity, i.e. the dispersion of identities within the village and identity and participation variables. The idea here is to test whether village-level heterogeneity indirectly affects such capital by influencing the effect of identity, and participation variables. While identity variables capture households' inherited level of social network capital, the participation variable captures an investment that household member may undertake, consciously or unconsciously, as a way to expand their personal network. Following the discussion on subjective identity in Section 6.2.3, we expect that heterogeneity affects the way individuals perceive their own identity and the identities of their neighbors, and thus their ability and willingness to connect with their neighbors. We present results in Column 3 of Table 6.4. As pointed out by Brambor et al. (2006), results displayed in regression tables provide limited information on the combined effect of two continuous variables, as the effect of one is estimated at the mean value of the other. We report in Figure 1, in addition to the table results, an estimation of the marginal effects of the three interacted variables for different values of ethnic heterogeneity with their respective 95% confidence interval (these are marginal effects on the log value of the dependent variable).

Results show that being a village native in a homogeneous village has a positive effect on households' access to credit from their social network, but that this effect decreases as ethnic heterogeneity in the village increases, and becomes quickly insignificant. No statistically significant effect is found regarding inherited social distance. Finally, the marginal effect of participation is also positively influenced by the degree of heterogeneity in the village. The marginal effect becomes significantly different from 0 for a value of ethnic fragmentation that is greater than 0.1.

Now we consider identity variables as indexes of households' innate social characteristics, and participation as an index of households' investment in social life. Results indicate here that while the innate characteristics pay off in homogeneous settings in terms of social network capital, they lose their value as the degree of heterogeneity increases. Conversely, while investment efforts have no significant return in homogeneous communities, their returns appear to increase as ethnic heterogeneity increases.

In the literature on individual social capital, differentiates distinction is often made between bonding and bridging social capital, the former referring to the strong links tying similar individuals (kinship ties and close friendship), while the latter refers to weaker ties linking more distant individuals (Gittell and Vidal, 1998; Woolcock, 1998; Woolcock and Narayan, 2000). The combination of the two, rather than the possession of one type, is seen as the most conducive to economic development (Granovetter, 1973; Woolcock and Narayan, 2000). The above-mentioned results show that bonding ties are more important in homogeneous settings, since identity plays a strong role there. A household's social network

Fig. 6.1: Estimated marginal effects of variables Native, Inherited social distance and Participation on social network capital, for different values of ethnic fragmentation

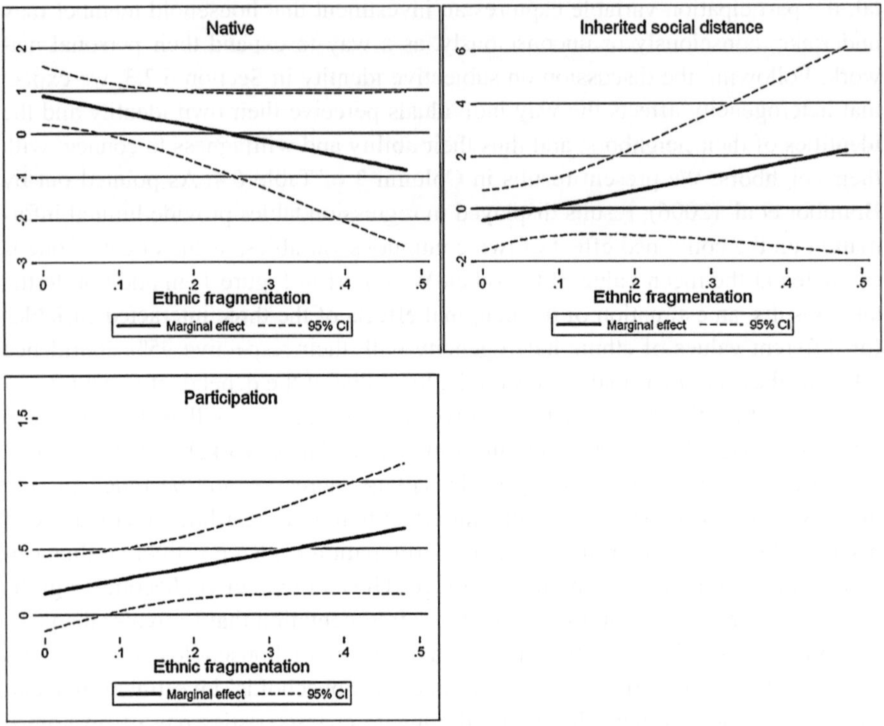

Marginal effects on the DV: log value of credit limit per unit of income.

is likely to be composed of households sharing the same identity, if this social network is sufficiently rich, individuals may not seek out additional relationships. This provides an explanation for the low return of participation on households' social network capital observed in homogeneous villages. As pointed out by Dasgupta et al. (2005), the marginal utility of creating new relationships, and therefore the incentives to do so, decreases as the amount of inherited relationships increases. In heterogeneous settings, on the other hand, where individuals are de facto more distant, individuals need to invest in expanding their networks, and participation in local organizations, as we saw, is a mean of doing so.

This raises questions regarding social mobility in homogeneous settings. If indeed social networks are determined by inherited position, the capacity of those who were born in low positions (or who migrated to the village) to move upward in the social and economic space appears to be rather limited. Heterogeneity instead seems to induce more flexibility in the way people connect with each other, and seems to favour the emergence of bridging connections. Hayami (2009) notes that the bridging connections help in adapting 'obsolete social norms' to current contexts. By 'obsolete social norms' he refers to norms and other institutions established overtime in traditional societies aimed at ensuring a social order, which may become inefficient as the economic and natural context evolves quickly (i.e. population growth and natural resource scarcity, transition to a market economy, etc.). These norms need considerable time to adapt to changing conditions, and the social interactions between individuals of different identities and traditions are likely to speed up this process. Based on our finding, we conclude that heterogeneous communities may be more likely to adapt rapidly to changing economic and ecological contexts.

6.6 Conclusion

The economic importance of ethnic diversity is gaining interest in the literature. Indicated as a factor that explains differences between levels of development observed in different countries or regions (Easterly and Levine, 1997), a growing body of literature seeks to explore the existence of mechanisms in place at the microeconomic level. Social capital, while difficult to capture in a single and tangible measure, appears as a powerful concept to explain differences in the economic performance of groups or societies. It is also often cited as one of the mechanisms through which ethnic heterogeneity undermines the economic performance. This article reviewed the literature and explored this question empirically. The existing empirical evidence on this relationship is relatively scarce, and findings appear to vary strongly from one country to another.

Recognizing the multi-faceted and complex nature of social capital, we use two measures and study two of its facets, namely participation in local organizations and households' social network capital, concentrating, for the latter, on a measure of households' financial social network. We make use of a particular setting in a rural district of Northern Vietnam in which ethnic heterogeneity within communities has been externally induced through resettlement policies conducted by the government in between 1960 an 1982. Using a detailed individual-, household- and village-level dataset collected in 2007/2008 on individuals' membership in organizations, households' credit access, and village-level settlement history among many other issues, we estimate econometrically the determinants of participation in local organizations as well as the determinants of households' social network capital.

Our results show that ethnic heterogeneity has an ambiguous impact on participation incentives, and that this effect strongly varies depending on the type of association considered, namely its political nature, the type of entry requirements, or whether it manages public or club goods. Using the past resettlement policy as an instrument, we are able to test and control for endogeneity of ethnic heterogeneity. The results do not confirm the negative relationship between ethnic diversity and participation in the management of public goods that is predicted in the literature, but show in fact a positive impact. Results, furthermore, indicate that competition for the management of these goods is important within villages, and that identity and ethnicity in particular play a significant role in determining who among village members will take part. We find however, a negative impact on participation incentives in the communist party, and also on professional organizations which are aimed at creating club goods. These results do not seem to confirm the prediction that participation in the elaboration of public goods decreases with the degree of heterogeneity.

We then develop an econometric model of household investment in social network capital, focusing on finance-related social networks. We hypothesize that households' level of such capital is both a function of innate characteristics, denoted by identity and inherited social distance between the household and its neighbours, and investment decisions such as participation in local organizations. We find no direct impact of ethnic heterogeneity on households' network social capital even after controlling for the endogeneity of this variable. However, a household's wealth level and variables denoting its identity and the participation of its members in local organizations appear to play a significant role. Furthermore, the test for the existence of interaction effects between ethnic heterogeneity, identity, and participation variables shows that ethnic heterogeneity has an indirect effect on households' social network capital. We find that households whose head or spouse are native to the village have a greater level of capital in homogeneous villages, but that this effect is diminished as ethnic heterogeneity increases. On the other hand, the effect of participation in local organizations is insignificant in the homogeneous setting, but becomes significant and increases in magnitude as the degree of heterogeneity increases. These results indicate that social mobility might be limited in homogeneous settings, where inherited economic statuses are likely to persist through generations. Conversely, in heterogeneous settings, households seem more disposed to invest in establishing bridging connections, which may help communities to adapt to rapidly changing conditions (population growth, climate change, etc.).

We believe that our approach and results, regarding the impact of ethnic diversity on participation and social networks capital, point to some understudied effects that deserve further attention. Ethnic heterogeneity matters economically, but its impact is ambiguous: sometimes positive and sometimes negative. We conclude that ethnically heterogeneous communities are not necessarily "bad" communities. On the contrary, results in this paper show that at the micro-level heterogeneity can also induce dynamism in social relations, favoring the establishment of bridging connections, which can help foster innovation, cross-cultural learning, social mobility and possibly economic development. These results are encouraging for overall development.

As previous works have shown, the institutional setting plays an important role in how individuals connect and cooperate with each other, and can compensate for village composition effects. Vietnam, after 40 years of collectivization, and about 15 years of a slow transition process, has maintained an important control over rural institutions. Everyday life and local behaviors still remain strongly influenced by communist precepts. For example, the government frequently delivers certificates rewarding 'good' farmers, neighbors, and others. Collective life and cooperation, in addition, are strongly anchored in Vietnamese culture. Although we cannot test the institutional hypothesis here due to lack of variability between villages, the institutional setting in place at the regional level may also play an important role. Finally, this paper has not addressed in detail the negative consequences induced by involuntary migrations and acculturation policies in Vietnam and its neighboring countries, such as Laos (Evrard and Goudineau, 2004). The cultural trauma and negative consequences on households' livelihoods remain largely understudied, although they are likely to be important. Therefore, the objective of this paper was by no means to promote or defend such policy, but rather to make use of this particular context to study the impact of ethnic heterogeneity.

Acknowledgement – The authors would like to thank Alwin Keil, Andreas Neef and Ling Yee Khor for valuable comments and helpful discussions on earlier versions of the paper. We also wish to thank Le Thi Ai Van, Dinh Thi Tuyet Van, Do Thi Nang, Susanne Ufer and Thea Nielsen for their precious collaboration in data collection, as well as Pham Thi My Dung from Hanoi University of Agriculture, and the Uplands Program for institutional and administrative support in Vietnam. Village heads and respondents are gratefully acknowledged for their time and willingness to take part in the survey. This research was funded by the DFG (Deutsche Forschungsgemeinschaft) through the special research program SFB 564. All errors remain ours.

6.7 Appendix

Table 6.5: Description and summary statistics of explanatory variables

Variable	Description	Var.a	Mean	S.D.	Min	Max
Age	Age of HH member	I	37.53	17.60	15	100
Male	=1 if member is male	I	0.48	0.50	0	1
Education	Education level of member	I	5.28	3.75	0	13
Farmer	=1 if main occupation is farming	I	0.77	0.42	0	1
Head/Spouse	=1 if member is the HH head or the spouse of the HH head	I	0.58	0.49	0	1
Age adults	Mean age of adults members	H	39.14	10.29	20.5	88
Share female	Share of females in HH	H	0.52	0.17	0	1
Education adults	Average education of adults in HH	H	4.93	2.91	0	12
Inherited social distance	Inherited social distance index	H	0.67	0.26	0.2	1.4
Native	=1 if HH head or spouse were born in village	H	0.79	0.41	0	1
Participation	Weighted mean of participation of all members above 15 in H. Weights account for degree of participation (highest for leader, lowest for non actives)	H	1.42	0.75	0	5
Farm size	Area operated in 2006 with long term use right, in acre per capita	H	27.80	19.37	0	132.9
Wealth index	Wealth index (Principal Component Factor)	H	0.00	1.00	–2.8	0.9
Elevation	Elevation of homestead in meter above the sea level (m.a.s.l.)	H	516.51	240.63	281	1088
HH size	Family size (all members > 6 month in the house)	H	4.64	1.99	1	15
Dependency ratio	Dependency ratio	H	0.33	0.23	0	1
Ethnic Thai	=1 if H head is Thai	H	0.75	0.43	0	1
Ethnic Kinh	=1 if H head is Kinh	H	0.09	0.29	0	1
Ethnic H'Mong	=1 if H head is H'Mong	H	0.15	0.35	0	1
Other ethnic	=1 if H head is from another ethnic group	H	0.01	0.10	0	1

Table 6.6: First stage instrumental equation of the participation model

DV: Ethnic fragmentation index	Coef.	t-stat
Age	0.035	(0.22)
Age square	–0.001	(0.61)
Male+	0.493	(0.80)
Education	–0.303	(1.64)
Speaks two languages+	0.619	(0.29)
Farmer+	–0.181	(0.16)
Head or spouse+	0.509	(0.46)
Household size	0.685	(1.22)
Dependency ratio	–1.582	(0.43)
Wealth index	0.295	(0.69)
Inherited social distance	14.181***	(4.22)
Native+	–2.640	(0.74)
Ethnic Kinh+[a]	5.242	(0.93)
Ethnic H'Mong+[a]	4.242	(0.96)
Elevation	–0.018***	(3.35)
Population	–0.004	(1.21)
Distance to paved road	0.149***	(4.14)
Distance to market	0.659***	(6.15)
Ethnic fragmentation	–	
Resettlement of households+	10.648***	(5.22)
N	986	
Adjusted-R^2	0.293	
F	10.560***	

t-stat are based on robust standard errors clustered at the household level
* p<0.10, ** p<0.05, *** p<0.01

7. Discussion and conclusions

Vietnam's upland regions have undergone important transformations during the last two decades. The transition of the country from a centrally-planned to a market-oriented system has entailed important institutional changes and a redefinition of the role of the state, markets and communities in allocating resources and organizing social and economic life.

While at the national level the reforms have permitted high economic growth rate and a drastic reduction of poverty, mountainous areas and its inhabitants, ethnic minorities, have lagged behind. In addition, the rapid population growth, combined with the intensification and expansion of agricultural production into fragile ecosystems have considerably increased pressure on natural resources, and accentuated the risk for the population to get trapped in vicious circle whereby poverty and environmental degradation mutually reinforce each other. Drawing on a conceptual framework that highlights the determinant role of institutions, in particular of land institutions, financial markets, and social capital in addressing the poverty-environment nexus, this thesis has investigated how the current institutional framework addresses challenges faced by mountainous people, and contributes to reduce poverty while enhancing environmental sustainability in the region.

In this perspective, this thesis first assessed the impact of Vietnam's land titling policy on tenure security, on the adoption of soil conservation technologies and on the functioning of land market. Second, it investigated the functioning of the credit market and the role and impact of the rural credit policy in this market. Finally, this research estimated the impact of ethnic heterogeneity on the formation of social capital, i.e. on participation in local organization and on social network capital. These analyses built on a primary database collected in Yen Chau district in 2007/2008 and on qualitative information collected through focus group discussions and semi-structured interviews with local stakeholders.

Extensive conclusions are provided at the end of each chapter (sections 3.7, 4.5, 5.7 and 6.6). In the following, we discuss the overall findings and their policy implications in section 7.1, and discuss orientation for further research in section 7.2.

7.1 Discussion of research findings and policy implications

This research has permitted the identification of successes and failures of the current institutional framework in addressing the challenges faced in mountainous regions. The results of this thesis and their implications are discussed under two perspectives: in the first one, we discuss how the combination of the state, market and communities in the uplands fosters or hampers sustainable development, and the second, more general, relates to the political economy of mountainous areas in Vietnam.

The four research chapters of this thesis provided insights into the interactions between the state, markets and communities in the uplands and their role in the rural economy. As put by Hayami (2001, p. 318), "*How to combine community, market and state in the economic system is probably the most important agenda in development economics*". The way these three organizations govern the economy and allocate resources determines the institutional framework and the incentive structure in the economy, and affects farmers' attitude towards resource use, and more generally the performance of the economy.

Research findings on land institutions highlights the existing frictions between the three organizations and tend to show in particular that the preponderance of the state limits the functioning of the two others. The state controls land use and land allocation and was able thanks to strong enforcement capacity to impose its law to communities and thereby weaken customary practices. The land market, on the other hand, does not function well as seen in chapter 4, and this is to a large extent explained by the persistent uncertainties on the content of users' rights, and the duration of those rights. We show in chapter 3 that the frequent reallocations conducted by the local government are an important source of uncertainty, even though the law is explicitly aimed at improving the functioning of the market.

The interplay between the state and the market in the rural credit sector is discussed in chapter 5. In this sector, and in the mountains particularly, the state prevents the emergence of independent formal financial institutions (i.e. from NGOs or independent banks), and thus the development of a competitive formal sector which could improve the functioning of the market, by fostering institutional innovations and improve credit access of poor households. The results show that the informal sector is well developed and competes to some extent with subsidized loans offered by both state-owned lenders. This demonstrates that conditions for the establishment of a competitive market are good in the area.

The strong enforcement and action capacities of the state certainly constitutes a strength. This has enabled the government to deeply transform the economy by involving communities, to increase its efficiency and to reduce poverty throughout the country (cf. introductory chapter 1). In comparison, the weakness of the state is perceived in many developing countries such as in Africa as a major problem. Yet, the preponderance of the state in many sectors of the economy impedes other organisations to function well. Considering the limitations inherent to the state, this imbalance is a source of inefficiency. First, inefficiencies in terms of poverty reduction are observed: the low poverty outreach of formal banks and the weakness of the land market disadvantage vulnerable and young households. These inefficiencies also affect the rural economy. The micro-credits are not used to finance profitable activities but permit instead well-off households to purchase durables. Second, the absence of formal saving options also increases the dependence towards credit of farmers and as a result the observed level of indebtedness is high in the area. The absence of a land market may also create efficiency losses, by preventing the transfer of land from less productive to more productive farmers, and by inducing young farmers to migrate. Finally, as shown in chapter 3, the situation on the land sector hampers environmental sustainability, and creates low incentives for conservation investments. State interventions such as on the credit market are very costly, and do not yield intended impacts. Whether more liberalized markets would lead to better outcomes could not be explored in this thesis. However, some observed feature on the informal credit sector, related to the presence of dense social fabric and the good level of interpersonal trust show that the conditions for the establishment of more liberalized institutions are quite good.

The strong control maintained by the state on the allocation of resources seems to be rather motivated by political reasons than being based on economic considerations. One pursued objective seems to maintain political order in the area, limit discontent among ethnic minorities, and keep control over populations that have opposed the state in the past[58]. Opposition between of ethnic minorities to state policies have occcured frequently during the last decades and still occured recently[59] (McElwee 2004).

58 Some ethnic groups, including H'Mong and some Thai subgroups joined the war against the Vietminh, on the side of the French (1950–1954) or the American (1962–1975).

59 In May 2011, thousands ethnic H'Mong people went down the street in Dien Bien Phu (located next to Son La, further West) to claim a greater autonomy. The demonstration was rapidly quashed by military intervention. The event was so rare that is was relayed by international media, see http://www.bbc.co.uk/news/world-asia-pacific-13284122 or http://www.nytimes.com/2011/05/06/world/asia/06briefs-Vietnam.html

The recent land allocation decisions in the area show the little value that the government attach to incentive-based mechanisms to address environmental challenges in the uplands. The response to soil degradation and related environment issues in Yen Chau has been to enforce the establishment of rubber plantations in the area, against farmers' will[60]. Yet, as shown in chapter 3, the clarification of the land law and a greater transparency in the state's objective could eliminate some suspicions regarding future reallocation and encourage the adoption of long term conservation technologies, at a lower cost. Experience elsewhere has shown that accounting for the environment while empowering local communities, politically and economically, is feasible through institutional innovations. Payment for Environmental Services (PES) mechanisms create incentives for land users (the suppliers of environmental services) to undertake conservation costs which are covered by identified beneficiaries ('buyers' of environmental services) (Engel et al., 2008).

Hence, one main recommendation of this thesis would be for the state to reinforce and/or develop institutions that support the functioning of incentive-based mechanisms in the area. Results show that the socio-economic conditions that are known to enable a good functioning of markets are quite good in the area. A greater transparency and credibility of the state is needed to achieve objectives of the land reform. The opening of the rural finance sector to independent actors (such as NGOs, or other independent MFIs) may help enhance the functioning of the credit market, particularly for the poor, and would cost less to the state. The implementation of PES systems, for instance in watersheds, between upland and lowland farmers, could help addressing environmental issues in the area[61].

Another important source of success of the current framework is to be found in the rich web of political and non-political organizations that exist in mountainous regions and that enable a dense social life, foster collective action, and

---

60 Several village in Yen Chau have been targeted to establish these plantations. This measure has met a strong opposition from the locals. In some cases, the police had to intervene and force farmers to plant the seedlings.

61 A study by Schad et al. (2011) studies the institutional system at place when a flood occurs in Yen Chau. It notably analyzes how this system affects farmers' perception of the event and their incentives to undertake preventive measure against floods (through soil conservation technologies for instance). Results show that the current system may not provide the right incentives to farmers, as responsibilities during flood management are not assigned relative to the role of actors in flooding, but follow hierarchical administrative structures

allow the government to reach even the most remote areas through policies. This is a legacy from the collectivization period during which rural life was organized by and around cooperatives. The Mass Organization which play an essential role in conveying government policy and in organizing farmers' everyday life are directly inherited from that period. The anthropological literature also stresses that the sense of collectivism is strongly anchored in the Vietnamese culture[62] and existed before the collectivization period. Authors in this field argue that this trait of the Vietnamese culture has been 'instrumented' by the communist regime to install the collectivist system, and explains why the regime has met little opposition during its implementation and persisted so long in Vietnam's rural areas (Jamieson, 1993, pp. 217, 256–257; Bergeret, 2003, p. 30–33; Kerkvliet, 2005, p. 39).

The Mass Organizations are present at all administrative level (from village to national government) and enable information to circulate vertically from the state to the grassroots and vice-versa, enabling the government to test and adjust policies based on feedbacks from the field. The essay of Kerkvliet (2005) shows concretely how lowland villages, through "everyday politics" have influenced national policy at the time of decollectivization. Hence, this model also empowers rural citizens.

However, mountain people and ethnic minorities, due to this minority status and their complex relationship with the state, remain politically marginalized in the sense that their voices are unlikely to weigh in the political[63].

The density of these social relationships also benefit the local economy, by reducing transaction costs through good levels of information circulation and the resulting high levels of trust[64]. Although this relation has not been tested empirically (because of the lack of variability in the data), this certainly explains the good performance of the informal credit sector (cf. the literature linking trust and the functioning of financial markets: Zeller, 1998; Karlan, 2005). As highlighted in

62 See Marr (2000) and Them (2006) for discussions of the notion of 'collectivism' and 'individualism' in Vietnamese culture.

63 See for instance McElwee (2004); Michaud (2000, 2009); Friederichsen and Neef (2010).

64 Respondents level of trust was measured through attitudinal survey questions. Data indicate very high level of trust, less than 3% of respondents only answered negatively to these trust related questions. These questions were adapted from the integrated questionnaire on social capital developed by the World Bank (Grootaert et al., 2002)

chapter 5, the difficulties met by formal institutions to increase their market share is explained by their own inefficiencies, but is also due to the presence of a vibrant informal sector.

This thesis has also investigated factors that enhance the formation of social capital within communities, by studying the determinants of two important of its facets, namely the participation of citizens in local organizations and the formation of financial social networks. The empirical literature, as highlighted in chapter 6, remains scarce on this issue. Results show that the ability (and willingness) for people to interact socially and form social networks does not only result from innate and given factors such as ethnicity and identity, but is also enhanced by the institutional and policy framework in place. Ethnic heterogeneity in addition is found to induce dynamism in these social relations, which contrasts with the relative 'conservatism' observed in homogeneous communities. The policy message delivered by these findings is not to be found in the resettlement policies conducted in Vietnam and elsewhere in South-East Asia that have induced potentially large negative human and cultural consequences, but rather in the positive consequences that social interactions among people from different cultural background can generate for the economy. Policies aimed at promoting social interactions, and by such favoring bridging connections among people of diverse cultural and social background may yield positive spillover effects in the economy.

One other recommendation is for Vietnam to support the emerging civil society in the mountains, to help it to organize politically and raise its voice. This will help the government to adapt policies to the context of the mountains, to better address the challenges and enhance sustainable development in the region. These measures may come as a complement to current policies, notably those aimed at providing safety nets and reducing poverty in the areas. Horizontal inequalities[65] between the Kinh and other ethnic groups remains a pressing issue in Vietnam (Minot et al., 2003; van de Walle and Gunewardena, 2001; Baulch et al., 2007).

As two recent reports from the World Bank have shown, Vietnam is making progress towards this direction, by enhancing the functioning of the state and

65 Horizontal inequality is defined by Stewart et al. (2005) as the inequalities between groups, while vertical inequality refers to the inequality between people. Horizontal inequality occurs when ethnicity – or other identity traits – is strongly correlated with economic status.

integrating young and qualified individuals in to the direction of the party, and by supporting the emergence of an independent civil society in the country (World Bank, 2009, 2004). These efforts will hopefully be pursued into mountainous regions and among ethnic minorities. The creation of the Committee for Ethnic Minorities (CEM, former Committee for Ethnic Minorities and Mountainous Areas (CEMMA)) in 1993 aimed representing the voice of ethnic minorities among public authorities, and its empowerment in 2008, demonstrates good intention in this direction, even though observers have judged its action to be rather inefficient (Baulch et al., 2007).

7.2 Orientation for further research

This thesis has provided substantial insights into the vast topic of the political and institutional economy of sustainable development of Vietnam's uplands. It was beyond the scope of this work to provide a comprehensive and exhaustive overview on this broad question. Instead this thesis has investigated three of its aspects which appeared during field work and in view of the literature, particularly relevant. Hence, many questions remain open for further research. First, possible extensions relate to issues that could not be investigated because of methodological limitations. Second the research has permitted to identify topics of high relevance which have not been much researched on in this thesis and in the literature in general.

One methodological constraint faced comes from the confinement of the data collection into one research district, and from the use of cross sectional data to estimate impacts. Both limitation in time and in space have limited the possibility to capture some variability that would have been needed to estimate their impacts[66]. Likewise, further investigations on the impact of land titling policy or on the impact of the credit policy could not be conducted, because of the lack of variability in variables of interest. For instance, in the credit case, it was not possible to construct a veritable control group composed of farmers who are similar to others but who do not have the possibility to borrow. Both formal banks operate indeed in all villages of the district. Thus, the expansion of existing data to

66 A methodological advantage however of such a database is that the heterogeneity in environmental and other unobservable aspects is limited and is unlikely to affect impact estimates.

a panel dataset where the time variability of variables of interest can be exploited would expand research possibilities to estimate impacts[67].

This thesis has not looked in detail at how the institutional framework at place affects the efficiency of land use in Yen Chau, and encourages in particular the adoption of new agricultural technologies. Yet, land tenure security, credit access and social networks are important factors of technological change (Feder et al., 1985). The agricultural extension institutions in place (Farmer Union or government agricultural extension) strongly promote the cultivation of maize as the most wealth-enhancing crop in the area. As shown by Keil et al. (2013) however, detrimental effects on the environment are important. Livestock farming in contrast appears as a potentially promising option to improve livelihoods in the short and long term. Further research may thus seek to investigate the institutional and economic constraints to the development of this sector.

Also, further investigation of the functioning of incentive-based mechanisms would provide valuable guidance to design future policies. In this vein, a deeper study of the land market, for instance through the econometric analysis of the determinants of farmers' engagement in transaction would provide valuable insights on the reasons underlying its malfunctioning and help to address this issue through policies. Similarly, research investigating the interactions between land and credit institutions would provide useful information relevant for policies. Finally, another perhaps more imminent research question would be to study potentials for the implementation of Payment for Environmental Services in the management of watersheds in Vietnam's mountainous areas as a way to break the poverty-environment nexus in the area.

Finally, this thesis has not looked in detail at the mechanisms underlying the persistence of inequalities between ethnic groups. Yet, these remain important issue, even in a small area as Yen Chau district. The H'Mong in particular remain significantly poorer than any of the other groups in Yen Chau. We could not identify in the data particular signs of discrimination in terms of land allocation, nor in the implementation of anti-poverty programs[68]. In all articles

---

67 This is actually one objective pursued in the fourth phase of F2 subproject.

68 Regarding credit however, we find that H'Mong villages receive significantly less VBSP funds than other villages. According to the data and information collected during informal discussions with H'Mong villagers and credit institutions, this was not always the case. But the non repayment of loans in these villages has induced VBSP to limit its action in those villages. One consequence is that H'Mong households heavily depend on unfair credit contracts delivered by shopkeepers established next to their village. Some get trapped in bad debt circle.

presented here, ethnic effects have been controlled for but were not studied in detail[69]. A deeper investigation of the constraints faced by H'Mong households to integrate the economy would help understanding mechanisms underlying the persistence of poverty in this group, and help identifying policy options to break this negative spiral.

69 The objective was to understand the economic mechanisms that cause these inequalities rather than to study the cultural and anthropological factors. A thorough analysis of the ethnic effects would have required a quasi-anthropological approach for which the author is not qualified.

Bibliography

ADB (2006). Country strategy and program: Vietnam 2007–2010. Technical report, Asian Development Bank, Manila, Philippines.

Ahlheim, M., O. Frör, A. Heinke, A. Keil, N. Duc, P. Van Dinh, C. Saint-Macary, and M. Zeller (2009). Landslides in mountainous regions of northern vietnam: Causes, protection strategies and the assessment of economic losses. *International Journal of Ecological Economics & Statistics 15*(F09), 20–33.

Ai, C. and E. C. Norton (2003). Interaction terms in logit and probit models. *Economics Letters 80*(1), 123–129.

Akerlof, G. A. (1970). The market for "lemons": quality uncertainty and the market mechanism. *Quarterly Journal of Economics 84* (3), 488–500.

Akerlof, G. A. and R. E. Kranton (2000). Economics and identity. *Quarterly Journal of Economics 115*(3), 715–753.

Akramov, K., B. Yu, and S. Fan (2010). Mountains, global food prices, and food security in the developing world. International Food Policy Research Institute, Washington D.C.

Alesina, A., R. Baqir, and W. Easterly (1999). Public goods and ethnic divisions. *Quarterly Journal of Economics 114*(4), 1243–1284.

Alesina, A. and E. La Ferrara (2000). Participation in heterogeneous communities. *The Quarterly Journal of Economics 115*(3), 847–904.

Alesina, A. and E. La Ferrara (2005). Ethnic diversity and economic performance. *Journal of Economic Literature 43*(3), 762–800.

Armendariz de Aghion, B. and J. Morduch (2007). *The Economics of Microfinance*. The MIT Press, Cambridge, Massachusetts.

Atwood, D. A. (1990). Land registration in Africa: the impact on agricultural production. *World Development 18*(5), 659–671.

Aubert, C., A. de Janvry, and E. Sadoulet (2009). Designing credit agent incentives to prevent mission drift in pro-poor microfinance institutions. *Journal of Development Economics 90*(1), 153–162.

Banerjee, A., E. Duflo, R. Glennerster, and C. Kinnan (2010). The miracle of microfinance? evidence from a randomized evaluation. mimeo.

Banerjee, A. V. (2003). Contracting constraints, credit markets and economic development. In M. Dewatripoint, L. P. Hansen, and S. J. Turnovsky (Eds.), *Advances in Economics and Econometrics: Theory and Applications, Eighth World Congress, Volume III*. Cambridge University Press.

Bardhan, P. (1989). Alternative approaches to the theory of instituions in economic development. In P. Bardhan (Ed.), *The Economic Theory of Agrarian Institutions*, Chapter 1, pp. 3–17. Oxford University Press, New York.

Barslund, M. and F. Tarp (2008). Formal and informal rural credit in four provinces of Vietnam. *Journal of Development Studies 44*(4), 485–503.

Bates, R. H. and I. Yachovlev (2002). Ethnicity, capital formation and conflict. In C. Grootaert and T. van Bastelaer (Eds.), *The Role of Social Capital in Development*. Cambridge, UK: Cambridge University Press.

Baulch, B., T. T. K. Chuyen, D. Haughton, and J. Haughton (2007). Ethnic minority development in Vietnam. *Journal of Development Studies 43*(7), 1151–1176.

Bell, C., T. N. Srintvasan, and C. Udry (1997). Rationing, spillover, and interlinking in credit markets: the case of rural Punjab. *Oxford Economic Papers 49*(4), 557–585.

Bergeret, P. (2003). *Paysans, État et Marchés au Vietnam: Dix ans de Coopération dans le Bassin du Fleuve Rouge*. Hommes et Sociétés. Paris, France: Editions du GRET, Editions Karthala.

Besley, T. (1994). How do market failures justify interventions in rural credit markets? *The World Bank Research Observer 9*(1), 27–47.

Besley, T. (1995). Property rights and investment incentives: theory and evidence from Ghana. *Journal of Political Economy 103*(5), 903–937.

Blundell, R. and M. Dias (2009). Alternative approaches to evaluation in empirical microeconomics. *Journal of Human Resources 44*(3), 565.

Blyth, S., B. Groombridge, I. Lysenko, L. Miles, and A. Newton (2002). *Mountain Watch: Environmental Change and Sustainable Development in Mountains*. Number 12. UNEP/Earthprint.

Boucher, S. and C. Guirkinger (2007). Risk, wealth, and sectoral choice in rural credit markets. *American Journal of Agricultural Economics 89*(4), 991–1004.

Boucher, S. R., B. L. Barham, and M. R. Carter (2005). The impact of "market-friendly" reforms on credit and land markets in Honduras and Nicaragua. *World Development 33*(1), 107–128.

Bourdieu, P. (1986). The forms of capital. In J. E. Richardson (Ed.), *Handbook of Theory of Research of the Sociology of Education*, pp. 46–58. Greenword Press.

Brambor, T., W. R. Clark, and M. Golder (2006). Understanding interaction models: improving empirical analyses. *Political Analysis 14*(1), 63–82.

Brasselle, A.-S., F. Gaspart, and J.-P. Platteau (2002). Land tenure security and investment incentives: puzzling evidence from Burkina Faso. *Journal of Development Economics 67*(2), 373–418.

Bromley, D. W. (2009). Formalising property relations in the developing world: The wrong prescription for the wrong malady. *Land Use Policy 26*(1), 20–27.

Bryant, J. (1998). Communism, poverty, and demographic change in North Vietnam. *Population and Development Review 24*(2), 235–269.

Caliendo, M. and S. Kopeinig (2008). Some practical guidance for the implementation of propensity score matching. *Journal of Economic Surveys 22*(1), 31–72.

Cameron, A. C. and P. K. Trivedi (2005). *Microeconometrics*. Cambridge University Press.

Carletto, C. and S. S. Morris (1999). Designing methods for the monitoring and evaluation of household food security rural development projects. International Food Policy Research Institute, Washington D. C.

Carter, M. R. and P. Olinto (2003). Getting institutions "right" for whom? credit constraints and the impact of property rights on the quantity and composition of investment. *American Journal of Agricultural Economics 85*(1), 173–186.

Cassar, A. and B. Wydick (2010). Does social capital matter? evidence from a five-country group lending experiment. *Oxford Economic Papers 62*(4), 715–739.

Chantarat, S. and C. Barrett (2011). Social network capital, economic mobility and poverty traps. *Journal of Economic Inequality 10*(3), 299–342.

Coleman, J. (1990). Social capital in the creation of human capital. *American Journal of Sociology 94* (Supplement), 59–120.

Conley, T. and C. Udry (2001). Social learning through networks: the adoption of new agricultural technology in Ghana. *American Journal of Agricultural Economics 83*(3), 668–673.

Conning, J. and M. Kevane (2002). Community-based targeting mechanisms for social safety nets: A critical review. *World Development 30*(3), 375–394.

Conning, J. H. and J. A. Robinson (2005). Property rights and the political organisation of agriculture. *Journal of Development Economics 82*(2), 416–447.

Conning, J. H. and C. Udry (2005). Rural financial markets in developing countries. In R. E. Evenson, P. Pingali, and T. P. Schultz (Eds.), *Handbook of Agricultural Economics, Vol. 3: Agricultural Development: Farmers, Farm Production and Farm Markets*. Elsevier Science.

Corlin, C. (2004). Hmong and the land question in Vietnam: National policy and local concepts of the environment. In N. Tapp, J. Michaud, C. Culas, and G. Y. Lee (Eds.), *Hmong/Miao in Asia*, pp. 295–320. Chiang Mai: Silkworm Books.

Cornia, G. A. and V. Popov (2001). Structural and institutional factors in the transition to the market economy: An overview. In G. A. Cornia and V. Popov (Eds.), *Transition and Institutions*, pp. 3–26. Oxford University Press.

Costa, D. L. and M. E. Kahn (2003). Civic Engagement and Community Heterogeneity: An Economist's Perspective. *Perspectives on Politics 1*(01), 103–111.

Cuong, N. V. (2008). Is a governmental micro-credit program for the poor really pro-poor? evidence from Vietnam. *The Developing Economies 46*(2), 151–187.

Dasgupta, P. (2000). Economic progress and the idea of social capital. In P. Dasgupta and I. Serageldin (Eds.), *Social Capital: A Multifaced Perspective*, pp. 325–424. The World Bank.

Dasgupta, P. (2003). Social capital and economic performance: analytics. In E. Ostrom and T. Ahn (Eds.), *Foundations of Social Capital*. Chelenham: Edward Elgar.

Dasgupta, P. and K.-G. Mäler (1995). Poverty, institutions, and the environmental resource-base. In J. Behrman and T. N. Srinivasan (Eds.), *Handbook of Development Economics*, Volume 3, Part 1, pp. 2371–2463. North-Holland: Elsevier.

Dasgupta, S., U. Deichmann, C. Meisner, and D. Wheeler (2005). Where is the poverty-environment nexus? evidence from Cambodia, Lao PDR, and Vietnam. *World Development 33*(4), 617–638.

de Janvry, A., M. Fafchamps, and E. Sadoulet (1991). Peasant household behaviour with missing markets: some paradoxes explained. *The Economic Journal 101*(409), 1400–1417.

de Soto, H. (2000). *The Mystery of Capital: Why Capitalism Triumphs in the West and Fails Everywhere Else?* New York: Basic Book.

De Weerdt, J. (2004). Risk-sharing and endogeneous network formation. In S. Dercon (Ed.), *Insurance Against Poverty*, pp. 197–216. Oxford University Press.

Deininger, K. (2003). *Land Policies for Growth and Poverty Reduction*. The World Bank and Oxford University Press.

Deininger, K. and G. Feder (2001). Land institutions and land markets. In B. Gardner and G. Rausser (Eds.), *Handbook of Agricultural Economics*, Volume 1, pp. 287–331. North-Holland: Elsevier.

Deininger, K. and S. Jin (2008). Land sales and rental markets in transition: Evidence from rural Vietnam. *Oxford Bulletin of Economics and Statistics 70*(1), 67–101.

Diagne, A. and M. Demont (2007). Taking a new look at empirical models of adoption: average treatment effect estimation of adoption rates and their determinants. *Agricultural Economics 37*(2–3), 201–210.

Diagne, A. and M. Zeller (2001). Access to credit and its impact in Malawi. International Food Policy Research Institute, Washington D.C.

Diagne, A., M. Zeller, and M. Sharma (2000). Empirical measurements of households' access to credit and credit constraints in developing countries:

methodological issues and evidence. International Food Policy Research Institute, Washington D.C.

DiPrete, T. and M. Gangl (2004). Assessing bias in the estimation of causal effects: Rosenbaum bounds on matching estimators and instrumental variables estimation with imperfect instruments. *Sociological Methodology 34*(1), 271–310.

Do, Q.-T. and L. Iyer (2003). Land rights and economic development: Evidence from Vietnam. The World Bank, Washington D.C.

Do, Q.-T. and L. Iyer (2008). Land titling and rural transition in Vietnam. *Economic Development and Cultural Change 56*(3), 531–579.

Dufhues, T. (2007). *Accessing Rural Finance: The Rural Financial Market in Northern Vietnam*. PhD thesis, University of Hohenheim.

Dufhues, T. and G. Buchenrieder (2005). Outreach of credit institutions and households' access constraints to formal credit in Northern Vietnam. Institute of Agricultural Economics and Social Sciences in the Tropics and Subtropics, Universtity of Hohenheim.

Dufhues, T., G. Buchenrieder, H. D. Quoc, and N. Munkung (2011). Social capital and loan repayment performance in southeast Asia. *Journal of Socio-Economics 40*(5), 679–691.

Duong, P. B. and Y. Izumida (2002). Rural development finance in Vietnam: a microeconometric analysis of household surveys. *World Development 30*(2), 319–335.

Duraiappah, A. K. (1998). Poverty and environmental degradation: A review and analysis of the nexus. *World Development 26*(12), 2169–2179.

Durlauf, S. N. and M. Fafchamps (2004). Social Capital. In P. Aghion and S. Durlauf (Eds.), *Handbook of Economic Growth*, Volume 1B. North-Holland: Elsevier.

Dutilly-Diané, C., E. Sadoulet, and A. De Janvry (2003). Household behavior under market failures: how natural resource management in agriculture promotes livestock production in the Sahel. *Journal of African Economies 2003* (12), 343–370.

Easterly, W. and R. Levine (1997). Africa's growth tragedy: Policies and ethnic divisions. *Quarterly Journal of Economics 112*(4), 1203–1250.

Engel, S., S. Pagiola, and S. Wunder (2008). Designing payments for environmental services in theory and practice: An overview of the issues. *Ecological Economics 65*(4), 663–674.

Epprecht, M., D. Müller, and N. Minot (2011). How remote are Vietnam's ethnic minorities? an analysis of spatial patterns of poverty and inequality. *The Annals of Regional Science 46*, 349–368.

Evrard, O. and Y. Goudineau (2004). Planned resettlement, unexpected migrations and cultural trauma in Laos. *Development and Change 35*(5), 937–962.

Fafchamps, M. and F. Gubert (2007). The formation of risk sharing networks. *Journal of Development Economics 83*(2), 326–350.

FAO (2011). Faostat, http://faostat.fao.org.

Feder, G. and N. Akihiko (1999). The benefits of land registration and titling: economic and social perspectives. *Land Use Policy 15*(1), 25–43.

Feder, G., R. E. Just, and D. Zilberman (1985). Adoption of agricultural innovations in developing countries: A Survey. *Economic Development and Cultural Change 33*(2), 255.

Feder, G. and R. Slade (1984). The acquisition of information and the adoption of new technology. *American Journal of Agricultural Economics 66*(3), 312–320.

Foster, A. D. and M. R. Rosenzweig (1995). Learning by doing and learning from others: human capital and technical change in agriculture. *Journal of Political Economy 103*(6), 1176–1209.

Friederichsen, R. J. and A. Neef (2010). Variations of late socialist development: Integration and marginalization in the Northern Uplands of Vietnam and Laos. *European Journal of Development Research 22*(4), 564–581.

Giné, X. (2011). Access to capital in rural Thailand: An estimated model of formal vs. informal credit. *Journal of Development Economics 96*(1), 16–29.

Gittell, R. and A. Vidal (1998). *Community Organizing: Building Social Capital as a Development Strategy*. Sage Publications.

Glaeser, E. L. (2001). The formation of social capital. *Canadian Journal of Policy Research 2*, 34–40.

Glaeser, E. L., D. Laibson, and B. Sacerdote (2002). An economic approach to social capital. *Economic Journal 112*(483), 437–458.

Glewwe, P., N. Agrawal, and D. Dollar (2004). *Economic Growth, Poverty, and Household Welfare in Vietnam*. The World Bank.

Goldstein, M., A. de Janvry, and E. Sadoulet (2005). Is a friend in need a friend indeed? inclusion and exclusion in mutual insurance networks in southern Ghana. In S. Dercon (Ed.), *Insurance against Poverty*, pp. 217–248. Oxford University Press.

Goldstein, M. and C. Udry (2008). The profits of power: land rights and agricultural investment in Ghana. *Journal of Political Economy 116*(6), 981–1022.

Granovetter, M. S. (1973). The strength of weak ties. *American Journal of Sociology 78*(6), 1360.

Grepperud, S. (1997). Soil conservation as an investment in land. *Journal of Development Economics 54*(2), 455–467.

Grootaert, C. and D. Narayan (2004). Local institutions, poverty and household welfare in Bolivia. *World Development 32*(7), 1179–1198.
Grootaert, C., D. Narayan, V. Nyhan Jones, and W. M. (2002). Measuring social capital – an integrated questionnaire. The World Bank.
Grosh, M. and P. Glewwe (2000). *Designing Household Survey Questionnaires for Developing Countries: Lessons from 15 Years of the Living Standards Measurement Study.* The World Bank.
GSO (2011). Statistical censuses and surveys http://www.gso.gov.vn/default_en.aspx.
Guirkinger, C. (2008). Understanding the coexistence of formal and informal credit markets in Piura, Peru. *World Development 36*(8), 1436–1452.
Habyarimana, J., M. Humphreys, D. N. Posner, and J. M. Weinstein (2007). Why does ethnic diversity undermine public goods provision? *American Political Science Review 101*, 709–725.
Haddad, L. and J. Maluccio (2003). Trust, membership in groups, and household welfare: evidence from KwaZulu-Natal, South Africa. *Economic Development and Cultural Change 51*(3), 573–601.
Hagos, F. and S. Holden (2006). Tenure security, resource poverty, public programs, and household plot-level conservation investments in the Highlands of Northern Ethiopia. *Agricultural Economics 34*(2), 183–196.
Haque, T. and L. Montesi (1996). Tenurial reforms and agricultural development in Vietnam. Food and Agriculture Organisation, Rome.
Hardy, A. (2000). Strategies of migration to upland areas in contemporary Vietnam. *Asia Pacific Viewpoint 41*(1), 23–34.
Hayami, Y. (2001). *Development Economics: From the Poverty to the Wealth of Nations Second Edition.* Oxford University Press.
Hayami, Y. (2009). Social capital, human capital and the community mechanism: toward a conceptual framework for economists. *Journal of Development Studies 45*(1), 96–123.
Hayes, J., M. Roth, and L. Zepeda (1997). Tenure security, investment and productivity in Gambian agriculture: a generalized probit analysis. *American Journal of Agricultural Economics 79*(2), 369–382.
Heckman, J. J. (1979). Sample selection bias as a specification error. *Econometrica 47*(1), 153–161.
Hoff, K. and J. Stiglitz (1990). Imperfect information and rural credit markets: Puzzles and policy perspectives. *The World Bank Economic Review 4*(3), 235–250.
Holden, S. T. and H. P. Binswanger (1998). Small-farmer decisionmaking, market imperfections and natural resource management in developing countries. In

E. Lutz (Ed.), *Agriculture and the Environment: Perspectives on Sustainable Rural Development*, pp. 50–70. The World Bank.

Holden, S. T., B. Shiferaw, and M. Wik (1998). Poverty, market imperfections and time preferences: of relevance for environmental policy? *Environment and Development Economics 3*(01), 105–130.

Isham, J. (2002). The effect of social capital on fertiliser adoption: evidence from rural Tanzania. *Journal of African Economies 11*(1), 39–60.

Jamieson, N., T. Le, and A. Rambo (1998). The development crisis in Vietnam's mountains. East-West Center, Honolulu, Hawai, USA.

Jamieson, N. L. (1993). *Understanding Vietnam*. University of California Press.

Karlan, D. (2005). Using experimental economics to measure social capital and predict financial decisions. *American Economic Review 95*(5), 1688–1699.

Karlan, D. and J. Morduch (2010). Access to finance. In D. Rodrik and M. Rosenzweig (Eds.), *Handbook of Development Economics*, Volume 5 of *Handbooks in Economics*, Chapter 71, pp. 4704–4777. North-Holland: Elsevier.

Keil, A., C. Saint-Macary, and M. Zeller (2013). Intensive commercial agriculture in fragile uplands of Vietnam : how to harness its poverty reduction potential while ensuring environmental sustainability. *Quaterly Journal of International Agriculture,* 52(1), 1–25.

Kerkvliet, B. J. T. (1995). Village-state relations in Vietnam: the effect of everyday politics on decollectivization. *Journal of Asian Studies 54*(2), 396–418.

Kerkvliet, B. J. T. (2005). *The Power of Everyday Politics: How Vietnamese Peasants Transformed National Policy*. New York, USA: Cornell University Press.

Khwaja, A. I. (2009). Can good projects succeed in bad communities? *Journal of Public Economics 93*(7–8), 899–916.

Kim Anh, B. (2010). The role of social network for credit access of rural poor households in northern Vietnam. Master's thesis, University of Hohenheim.

Kirk, M. and N. Tuan (2009). Land-tenure policy reforms: Decollectivization and the Doi Moi system in Vietnam. International Food Policy Research Institute, Washington D.C.

Kochar, A. (1997). An empirical investigation of rationing constraints in rural credit markets in India. *Journal of Development Economics 53*(2), 339–371.

La Ferrara, E. (2002). Inequality and group participation: theory and evidence from rural Tanzania. *Journal of Public Economics 85*(2), 235–273.

Lapenu, C. (2000). The role of the state in promoting microfinance institutions. International Food Policy Research Institute,Washington D.C.

Lundgren, B. (1982). Introduction. *Agroforestry Systems 1*(1), 1–4.

Lutz, E., S. Pagiola, and C. Reiche (1994). Lessons from economic and institutional analyses of soil conservation projects in Central America and the Caribbean. In E. Lutz, S. Pagiola, and C. Reiche (Eds.), *Economic and Institutional Analyses of Soil Conservation Projects in Central America and the Caribbean*, The World Bank Environment Paper, pp. 3–18. The World Bank, Washington D.C.

Macours, K. and J. F. M. Swinnen (2002). Patterns of agrarian transition. *Economic Development and Cultural Change 50*(2), pp. 365–394.

Mansuri, G. and V. Rao (2004). Community-based and -driven development: A critical review. *The World Bank Research Observer 19*(1), 1–39.

Marr, D. (2000). Concepts of 'individualism' and 'self' in twentieth-century Vietnam. *Modern Asian Studies 34*(4), 769–796.

McConnell, K. (1983). An economic model of soil conservation. *American Journal of Agricultural Economics 65*(1), 83–89.

McElwee, P. (2004). Becoming socialist or becoming Kinh? In C. R. Duncan (Ed.), *Civilizing the Margins: Southeast Asian Government Policies for the Development of Minorities*. Ithaca, New York: Cornell University Press.

Meinzen-Dick, R. and A. Knox (2001). Collective action, property rights, and devolution of natural resource management: exchange of knowledge and implications for policy. CAPRi Working Paper No.11.

Meinzen-Dick, R. and E. Mwangi (2009). Cutting the web of interests: Pitfalls of formalizing property rights. *Land Use Policy 26*(1), 36–43.

Meinzen-Dick, R. and R. Pradhan (2003). Legal pluralism and dynamic property rights. International Food Policy Research Institute, Washington, D.C.

Mellac, M. (2006). From "Muong ricefield" to "Cooperative ricefield": some thoughts on the evolution of the cognitive categories used by the Tai in northwest Vietnam. In *International symposium: "A la frontière des questions foncières – At the frontier of land issues", May 2006, Montpellier, France.*

Menkhoff, L. and O. Rungruxsirivorn (2011). Do village funds improve access to finance? evidence from Thailand. *World Development 39*(1), 110–122.

Michaud, J. (2000). The Montagnards and the state in Northern Vietnam from 1802 to 1975: a historical overview. *Ethnohistory 47*(2), 333–368.

Michaud, J. (2009). Handling mountain minorities in China, Vietnam and Laos: from history to current concerns. *Asian Ethnicity 10*(1), 25–49.

Michaud, J., S. Turner, and Y. Roche (2002). Mapping ethnic diversity in highland Northern Vietnam. *GeoJournal 57*, 305–323.

Miguel, E. and M. K. Gugerty (2005). Ethnic diversity, social sanctions, and public goods in Kenya. *Journal of Public Economics 89*(11–12), 2325–2368.

Minot, N., B. Baulch, and M. Epprecht (2003). Poverty and inequality in Vietnam: Spatial patterns and geographic determinants. International Food Policy Research Institute, Washington, D.C.

MIX (2011). Microfinance information exchange database http://www.mixmarket.org.

Mogues, T. and M. R. Carter (2005). Social capital and the reproduction of economic inequality in polarized societies. *Journal of Economic Inequality 3*(3), 193–219.

Mohieldin, M. S. and P. W. Wright (2000). Formal and informal credit markets in Egypt. *Economic Development and Cultural Change 48*(3), pp. 657–670.

Montes, M. F. (2001). Vietnam: transition as a Socialist Project in East Asia. In G. A. Cornia and V. Popov (Eds.), *Transition and Institutions*. New York: Oxford University Press.

Narayan, D. and L. Pritchett (1999). Cents and sociability: household income and social capital in rural Tanzania. *Economic Development and Cultural Change 47*(4), 871–897.

Neef, A., R. J. Friederichsen, C. Sangkapitux, and N. T. Thac (2002). Sustainable livelihoods in mountainous regions of northern Vietnam: from technology-oriented to people centered concepts. Paper presented at the Montane Mainland Southeast Asia (MMSEA) III Conference, 25–28 August 2002, Lijiang, China.

North, D. C. (1990). *Institutions, Institutional Change and Economic Performance*. London: Cambridge University Press.

Okten, C. and U. O. Osili (2004). Social networks and credit access in Indonesia. *World Development 32*(7), 1225–1246.

Ostrom, E. (2000). Collective action and the evolution of social norms. *The Journal of Economic Perspectives 14*(3), 137–158.

Otsuka, K. (2007). Efficiency and equity effects of land markets. In R. Evenson and P. Pingali (Eds.), *Handbook of Agricultural Economics*, Volume 3, pp. 2671–2703. Elsevier.

Pande, R. and C. Udry (2006). Institutions and development: A view from below. In *Advances in Economics and Econometrics: Theory and Applications, Ninth World Congress*, Volume 2, pp. 349–411.

Pelley, P. M. (2002). *Postcolonial Vietnam – New histories of the national past*. Asia Pacific: Culture, Politics, and Society. Durham and London: Duke University Press.

Pender, J. and M. Fafchamps (2001). Land lease markets and agricultural efficiency in Ethiopia. International Food Policy Research Institute, Department of economics University of Oxford, mimeo.

Pender, J. L. (1996). Discount rates and credit markets: theory and evidence from rural India. *Journal of Development Economics 50*(2), 257–296.
Petrick, M. (2005). Empirical measurement of credit rationing in agriculture: a methodological survey. *Agricultural Economics 33*(2), 191–203.
Place, F. and B. Swallow (2000). Assessing the relationships between property rights and technology adoption in smallholder agriculture: a review of issues and empirical methods. International Food Policy Research Institute, Washington D.C.
Platteau, J.-P. (1996). The Evolutionary Theory of Land Rights as Applied to SubSaharan Africa: A Critical Assessment. *Development and Change 27*(1), 29–86.
Platteau, J.-P. (2000). *Institutions, Social Norms and Economic Development.* Harwood Academic Publishers.
Platteau, J.-P. and A. Abraham (2002). Participatory development in the presence of endogenous community imperfections. *Journal of Development Studies 39*(2), 104–136.
Putnam, R. D. (1993). *Making Democracy Work: Civic Traditions in Modern Italy*. New Jersey: Princeton University Press.
Putnam, R. D. (1995). Tuning in, tuning out: the strange disappearance of social capital in america. *PS: Political Science and Politics 28*(4), 664–683.
Putnam, R. D. (2007). E Pluribus Unum: diversity and community in the twenty-first Century. *Scandinavian Political Studies 30*(2), 137–174.
Que, T. T. (2005). Land and agricultural land management in Vietnam. Technical report, Institute of Development Economies, Japan External Trade Organisation.
Ravallion, M. and D. van de Walle (2004). Breaking up the collective farms. *Economics of Transition 12*(2), 201–236.
Ravallion, M. and D. van de Walle (2008a). Does rising landlessness signal success or failure for Vietnam's agrarian transition? *Journal of Development Economics 87*(2), 191–209.
Ravallion, M. and D. van de Walle (2008b). *Land in Transition: Reform and Poverty in Rural Vietnam*. New York: Palgrave MacMillan.
Ray, D. (1998). *Development Economics*. Princeton, New Jersey: Princeton University Press.
Reardon, T. and S. A. Vosti (1995). Links between rural poverty and the environment in developing countries: Asset categories and investment poverty. *World Development 23*(9), 1495–1506.
Rivers, D. and Q. H. Vuong (1988). Limited information estimators and exogeneity tests for simultaneous probit models. *Journal of Econometrics 39*(3), 347–366.

Rosenbaum, P. (2002). *Observational studies*. Springer Verlag.

Rosenbaum, P. and D. Rubin (1983). The central role of the propensity score in observational studies for causal effects. *Biometrika 70*(1), 41.

Rozelle, S. and J. F. M. Swinnen (2004). Success and failure of reform: insights from the transition of agriculture. *Journal of Economic Literature 42*(2), 404–456.

Rutherford, S. (2000). *The poor and their money*. Oxford University Press New Delhi.

Saint-Macary, C., A. Keil, M. Zeller, F. Heidhues, and P. T. M. Dung (2010). Land titling policy and the adoption of soil conservation technologies in the uplands of Northern Vietnam. *Land Use Policy 27*(4), 617–627.

Santos, P. and C. B. Barrett (2010). Identity, interest and information search in a dynamic rural economy. *World Development 38*(12), 1788 – 1796.

Schad, I., P. Schmitter, C. Saint-Macary, A. Neef, M. Lamers, L. Nguyen, T. Hilger, and V. Hoffman (2012). Why do people not learn from flood disasters? evidence from Vietnam's northwestern mountains. *Natural Hazards*, 2012, 66(2), p. 221–241.

Scherr, S. J. (2000). A downward spiral? research evidence on the relationship between poverty and natural resource degradation. *Food Policy 25*(4), 479–498.

Schuller, T., S. Baron, and J. Field (2000). Social capital: A review and critique. In S. Baron, J. Field, and T. Schuller (Eds.), *Social Capital: Critical Perspectives*, pp. 1–38. Oxford University Press.

Scoones, I. (1998). Sustainable rural livelihoods: a framework for analysis. Institute for Development Studies, Brighton, England.

Shively, G. E. (2001). Poverty, consumption risk and soil conservation. *Journal of Development Economics 65*(2), 267–290.

Siamwalla, A., C. Pinthong, N. Poapongsakorn, P. Satsanguan, P. Nettayarak, W. Mingmaneenakin, and Y. Tubpun (1990). The Thai rural credit system: Public subsidies, private information, and segmented markets. *The World Bank Economic Review 4*(3), 271–295.

Sikor, T. (2004). Conflicting concepts: contested land relations in North-western Vietnam. *Conservation and Society 2*(1), 75–95.

Sobel, J. (2002). Can we trust social capital? *Journal of Economic Literature 40*(1), 139–154.

Socialist Republic of Vietnam (2004). Land law. National Political Publisher, Hanoi, Vietnam.

Soule, M. J., A. Tegene, and K. D. Wiebe (2000). Land tenure and the adoption of conservation practices. *American Journal of Agricultural Economics 84*(4), 993–1005.

Stewart, F., G. Brown, and L. Mancini (2005). Why horizontal inequalities matter: some implications for measurement. CRISE Working Paper No. 19.

Stiglitz, J. E. (1989). Rational peasants, efficient institutions, and a theory of rural organization: Methodological remarks for development economics. In P. Bardhan (Ed.), *The Economic Theory of Agrarian Institutions*, Chapter 2, pp. 18–29. Oxford University Press, New York, USA.

Stiglitz, J. E. and A. Weiss (1981). Credit rationing in markets with imperfect information. *American Economic Review 71*(3), 393–410.

Them, T. N. (2006). *Recherche sur l'Identité de la Culture Vietnamienne*. Hanoi: The Gioi Publisher.

Tra Pham, T. T. and R. Lensink (2007). Lending policies of informal, formal and semiformal lenders. *Economics of Transition 15*(2), 181–209.

Udry, C. and T. Conley (2004). Social networks in Ghana. Yale University, Center for Economic Growth, Center Discussion Paper No. 888.

van de Walle, D. and D. Gunewardena (2001). Sources of ethnic inequality in Viet Nam. *Journal of Development Economics 65*(1), 177–207.

Varughese, G. and E. Ostrom (2001). The contested role of heterogeneity in collective action: some evidence from community forestry in Nepal. *World Development 29*(5), 747–765.

Vosti, S. A. and T. Reardon (1997). *Sustainability, Growth and Poverty Alleviation: A Policy and Agroecological Perspective*. Johns Hopkins University Press, Baltimore, MD.

Williamson, O. (1979). Transaction-cost economics: the governance of contractual relations. *Journal of Law and Economics 22*(2), 233–261.

Wirth, T., D. C. Thu, and A. Neef (2004). Traditional land tenure among the Black Thai and its implication on the land allocation in Yen Chau district, Son La province, Northwest Vietnam. In G. Gerold, M. Fremerey, and E. Guhardja (Eds.), *Land use, nature conservation and the stability of rainforest margins in southeast Asia*, pp. 119–134. Berlin, Heidelberg, New York, London, Paris and Tokyo: Springer-Verlag.

Woolcock, M. (1998). Social capital and economic development: Toward a theoretical synthesis and policy framework. *Theory and Society 27*(2), 151–208.

Woolcock, M. and D. Narayan (2000). Social capital: implications for development theory, research, and policy. *The World Bank Research Observer 15*(2), 225–249.

Wooldridge, J. M. (2002a). *Econometric analysis of cross section and panel data*. The MIT Press, London U.K.

Wooldridge, J. M. (2002b). *Introductory Econometrics: A Modern Approach, 2nd Edition*. South-Western College.
Wooldridge, J. M. (2006). Cluster-sample methods in applied econometrics: an extended analysis. Department of economics, Michigan State University.
World Bank (1991). World Bank Development Report 1992: Development and the Environment. The World Bank, Washington D.C.
World Bank (2003). Vietnam development report 2004: Poverty. The World Bank, Washington D.C., and Hanoi, Vietnam.
World Bank (2004). Vietnam Development Report 2005: Governance. The World Bank, Washington D.C., and Hanoi, Vietnam.
World Bank (2006). Vietnam development report 2007: Aiming High. The World Bank, Washington D.C., and Hanoi, Vietnam.
World Bank (2008). Vietnam Development Report 2009: Capital Matters. The World Bank, Washington D.C., and Hanoi, Vietnam.
World Bank (2009). Vietnam Development Report 2010: Modern Institutions. The World Bank, Washington D.C., and Hanoi, Vietnam.
World Bank (2011). World development indicators, http://data.worldbank.org/indicator.
Zeller, M. (1994). Determinants of credit rationing: a study of informal lenders and formal credit groups in Madagascar. *World Development 22*(12), 1895–1907.
Zeller, M. (1998). Determinants of repayment performance in credit groups: The role of program design, intragroup risk pooling, and social cohesion. *Economic Development and Cultural Change 46*(3), 599–620.
Zeller, M. (2003). Models of rural financial institutions. In *Paving the Way Forward for Rural Finance Conference, June 2003, Washington D.C.*, Volume 2.
Zeller, M., T. Beuchelt, I. Fischer, and F. Heidhues (2010). Linkages between poverty and sustainable agricultural and rural development in the uplands of southeast Asia. In T. Tscharntke, C. Leuschner, E. Veldkamp, H. Faust, E. Guhardja, and A. Bidin (Eds.), *Tropical Rainforests and Agroforests under Global Change*, Environmental Science and Engineering, pp. 511–527. Springer Berlin Heidelberg.
Zeller, M. and R. Meyer (2002). *The triangle of microfinance: Financial sustainability, outreach, and impact*. The Johns Hopkins University Press.
Zeller, M. and M. Sharma (2000). Many borrow, more save, and all insure: implications for food and micro-finance policy. *Food Policy 25*(2), 143–167.

Development Economics and Policy

Series edited by Franz Heidhues, Joachim von Braun and Manfred Zeller

Band 1 Andrea Fadani: Agricultural Price Policy and Export and Food Production in Cameroon. A Farming Systems Analysis of Pricing Policies. The Case of Coffee-Based Farming Systems. 1999.

Band 2 Heike Michelsen: Auswirkungen der Währungsunion auf den Strukturanpassungsprozeß der Länder der afrikanischen Franc-Zone. 1995.

Band 3 Stephan Bea: Direktinvestitionen in Entwicklungsländern. Auswirkungen von Stabilisierungsmaßnahmen und Strukturreformen in Mexiko. 1995.

Band 4 Franz Heidhues / François Kamajou: Agricultural Policy Analysis – Proceedings of an International Seminar, held at the University of Dschang, Cameroon on May 26 and 27 1994, funded by the European Union under the Science and Technology Program (STD). 1996.

Band 5 Elke M. Förster: Protection or Liberalization? A Policy Analysis of the Korean Beef Sector. 1996.

Band 6 Gertrud Schrieder: The Role of Rural Finance for Food Security of the Poor in Cameroon. 1996.

Band 7 Nestor R. Ahoyo Adjovi: Economie des Systèmes de Production intégrant la Culture de Riz au Sud du Bénin: Potentialités, Contraintes et Perspectives. 1996.

Band 8 Jenny Müller: Income Distribution in the Agricultural Sector of Thailand. Empirical Analysis and Policy Options. 1996.

Band 9 Michael Brüntrup: Agricultural Price Policy and its Impact on Production, Income, Employment and the Adoption of Innovations. A Farming Systems Based Analysis of Cotton Policy in Northern Benin. 1997.

Band 10 Justin Bomda: Déterminants de l'Epargne et du Crédit, et leurs Implications pour le Développement du Système Financier Rural au Cameroun. 1998.

Band 11 John M. Msuya: Nutrition Improvement Projects in Tanzania: Implementation, Determinants of Performance, and Policy Implications. 1998.

Band 12 Andreas Neef: Auswirkungen von Bodenrechtswandel auf Ressourcennutzung und wirtschaftliches Verhalten von Kleinbauern in Niger und Benin. 1999.

Band 13 Susanna Wolf (ed.): The Future of EU-ACP Relations. 1999.

Band 14 Franz Heidhues / Gertrud Schrieder (eds.): Romania – Rural Finance in Transition Economies. 2000.

Band 15 Katinka Weinberger: Women's Participation. An Economic Analysis in Rural Chad and Pakistan. 2000.

Band 16 Christof Batzlen: Migration and Economic Development. Remittances and Investments in South Asia: A Case Study of Pakistan. 2000.

Band 17 Matin Qaim: Potential Impacts of Crop Biotechnology in Developing Countries. 2000.

Band 18 Jean Senahoun: Programmes d'ajustement structurel, sécurité alimentaire et durabilité agricole. Une approche d'analyse intégrée, appliquée au Bénin. 2001.

Band 19 Torsten Feldbrügge: Economics of Emergency Relief Management in Developing Countries. With Case Studies on Food Relief in Angola and Mozambique. 2001.

Band 20 Claudia Ringler: Optimal Allocation and Use of Water Resources in the Mekong River Basin: Multi-Country and Intersectoral Analyses. 2001.

Band 21 Arnim Kuhn: Handelskosten und regionale (Des-)Integration. Russlands Agrarmärkte in der Transformation. 2001.

Band 22 Ortrun Anne Gronski: Stock Markets and Economic Growth. Evidence from South Africa. 2001.

Band 23 Patrick Webb / Katinka Weinberger (eds.): Women Farmers. Enhancing Rights, Recognition and Productivity. 2001.

Band 24 Mingzhi Sheng: Lebensmittelkonsum und -konsumtrends in China. Eine empirische Analyse auf der Basis ökonometrischer Nachfragemodelle. 2002.

Band 25 Maria Iskandarani: Economics of Household Water Security in Jordan. 2002.

Band 26 Romeo Bertolini: Telecommunication Services in Sub-Saharan Africa. An Analysis of Access and Use in the Southern Volta Region in Ghana. 2002.

Band 27 Dietrich Müller-Falcke: Use and Impact of Information and Communication Technologies in Developing Countries' Small Businesses. Evidence from Indian Small Scale Industry. 2002.

Band 28 Wolfram Erhardt: Financial Markets for Small Enterprises in Urban and Rural Northern Thailand. Empirical Analysis on the Demand for and Supply of Financial Services, with Particular Emphasis on the Determinants of Credit Access and Borrower Transaction Costs. 2002.

Band 29 Wensheng Wang: The Impact of Information and Communication Technologies on Farm Households in China. 2002.

Band 30 Shyamal K. Chowdhury: Institutional and Welfare Aspects of the Provision and Use of Information and Communication Technologies in the Rural Areas of Bangladesh and Peru. 2002.

Band 31 Annette Luibrand: Transition in Vietnam. Impact of the Rural Reform Process on an Ethnic Minority. 2002.

Band 32 Felix Ankomah Asante: Economic Analysis of Decentralisation in Rural Ghana. 2003.

Band 33 Chodechai Suwanaporn: Determinants of Bank Lending in Thailand: An Empirical Examination for the Years 1992 to 1996. 2003.

Band 34 Abay Asfaw: Costs of Illness, Demand for Medical Care, and the Prospect of Community Health Insurance Schemes in the Rural Areas of Ethiopia. 2003.

Band 35 Gi-Soon Song: The Impact of Information and Communication Technologies (ICTs) on Rural Households. A Holistic Approach Applied to the Case of Lao People's Democratic Re- public. 2003.

Band 36 Daniela Lohlein: An Economic Analysis of Public Good Provision in Rural Russia. The Case of Education and Health Care. 2003.

Band 37 Johannes Woelcke. Bio-Economics of Sustainable Land Management in Uganda. 2003.

Band 38 Susanne M. Ziemek: The Economics of Volunteer Labor Supply. An Application to Countries of a Different Development Level. 2003.

Band 39 Doris Wiesmann: An International Nutrition Index. Concept and Analyses of Food Insecurity and Undernutrition at Country Levels. 2004.

Band 40 Isaac Osei-Akoto: The Economics of Rural Health Insurance. The Effects of Formal and Informal Risk-Sharing Schemes in Ghana. 2004.

Band 41 Yuansheng Jiang: Health Insurance Demand and Health Risk Management in Rural China. 2004.

Band 42 Roukayatou Zimmermann: Biotechnology and Value-added Traits in Food Crops: Relevance for Developing Countries and Economic Analyses. 2004.

Band 43 F. Markus Kaiser: Incentives in Community-based Health Insurance Schemes. 2004.

Band 44 Thomas Herzfeld: *Corruption begets Corruption*. Zur Dynamik und Persistenz der Korruption. 2004.

Band 45 Edilegnaw Wale Zegeye: The Economics of On-Farm Conservation of Crop Diversity in Ethiopia: Incentives, Attribute Preferences and Opportunity Costs of Maintaining Local Varieties of Crops. 2004.

Band 46 Adama Konseiga: Regional Integration Beyond the Traditional Trade Benefits: Labor Mobility contribution. The Case of Burkina Faso and Côte d'Ivoire. 2005.

Band 47 Beyene Tadesse Ferenji: The Impact of Policy Reform and Institutional Transformation on Agricultural Performance. An Economic Study of Ethiopian Agriculture. 2005.

Band 48 Sabine Daude: Agricultural Trade Liberalization in the WTO and Its Poverty Implications. A Study of Rural Households in Northern Vietnam. 2005.

Band 49 Kadir Osman Gyasi: Determinants of Success of Collective Action on Local Commons. An Empirical Analysis of Community-Based Irrigation Management in Northern Ghana. 2005.

Band 50 Borbala E. Balint: Determinants of Commercial Orientation and Sustainability of Agricultural Production of the Individual Farms in Romania. 2006.

Band 51 Pamela Marinda: Effects of Gender Inequality in Resource Ownership and Access on Household Welfare and Food Security in Kenya. A Case Study of West Pokot District. 2006.

Band 52 Charles Palmer: The Outcomes and their Determinants from Community-Company Contracting over Forest Use in Post-Decentralization Indonesia. 2006.

Band 53 Hardwick Tchale: Agricultural Policy and Soil Fertility Management in the Maize-based Smallholder Farming System in Malawi. 2006.

Band 54 John Kedi Mduma: Rural Off-Farm Employment and its Effects on Adoption of Labor Intensive Soil Conserving Measures in Tanzania. 2006.

Band 55 Mareike Meyn: The Impact of EU Free Trade Agreements on Economic Development and Regional Integration in Southern Africa. The Example of EU-SACU Trade Relations. 2006.

Band 56 Clemens Breisinger: Modelling Infrastructure Investments, Growth and Poverty Impact. A Two-Region Computable General Equilibrium Perspective on Vietnam. 2006.

Band 57 Meike Wollni: Coping with the Coffee Crisis. An Analysis of the Production and Marketing Performance of Coffee Farmers in Costa Rica. 2007.

Band 58 Franklin Simtowe: Performance and Impact of Microfinance. Evidence from Joint Liability Lending Programs in Malawi. 2008.

Band 59 Xiangping Jia: Credit Rationing and Institutional Constraint. Evidence from Rural China. 2008.

Band 60 Holger Seebens: The Economics of Gender and the Household in Developing Countries. 2009.

Band 61 Stephan Piotrowski: Land Property Rights and Natural Resource Use. An Analysis of Household Behavior in Rural China. 2009.

Band 62 Sebastian M. Scholz: Rural Development through Carbon Finance. Forestry Projects under the Clean Development Mechanism of the Kyoto Protocol. Assessing Smallholder Participation by Structural Equation Modeling. 2009.

Band 63 Jakob Rupert Friederichsen: Opening Up Knowledge Production through Participatory Research? Agricultural Research for Vietnam's Northern Uplands. 2009.

Band 64 Olivier Ecker: Economics of Micronutrient Malnutrition. The Demand for Nutrients in Sub-Saharan Africa. 2009.

Band 65 Julia Johannsen: Operational Assessment of Monetary Poverty by Proxy Means Tests. 2009

Band 66 Ephraim Nkonya / Nicolas Gerber / Philipp Baumgartner / Joachim von Braun / Alessandro De Pinto / Valerie Graw / Edward Kato / Julia Kloos / Teresa Walter: The Economics of Land Degradation. Toward an Integrated Global Assessment. 2011.

Band 67 S. Idriss Nazaire Houssou: Operational Poverty Targeting by Proxy Means Tests. Models and Policy Simulations for Malawi. 2013.

Band 68 Abdul Salam Lodhi: Education, Child Labor and Human Capital Formation in Selected Urban and Rural Settings of Pakistan. 2013.

Band 69 Evita Hanie Pangaribowo: Household Food Consumption, Women´s Asset and Food Policy in Indonesia. 2013.

Band 70 Dan Liu: China's New Rural Cooperative Medical Scheme. Evolution, Design and Impacts. 2013.

Band 71 Camille Saint-Macary: Microeconomic Impacts of Institutional Change in Vietnam's Northern Uplands. Empirical Studies on Social Capital, Land and Credit Institutions. 2014.

www.peterlang.com